MATHEMATICS, METROLOGY, AND MODEL CONTRACTS

A Codex from Late Antique Business Education

(*P.Math.*)

ISAW Monographs

ISAW Monographs publishes authoritative studies of new evidence and research into the texts, archaeology, art history, material culture, and history of the cultures and periods representing the core areas of study at NYU's Institute for the Study of the Ancient World. The topics and approaches of the volumes in this series reflect the intellectual mission of ISAW as a center for advanced scholarly research and graduate education whose aim is to encourage the study of the economic, religious, political, and cultural connections between ancient civilizations, from the Western Mediterranean across the Near East and Central Asia, to China.

MATHEMATICS, METROLOGY, AND MODEL CONTRACTS
A Codex from Late Antique Business Education
(*P.Math.*)

EDITED BY

ROGER S. BAGNALL AND ALEXANDER JONES

WITH THE ASSISTANCE OF

KATJA KOSOVA AND ZOE MISIEWICZ

AND WITH CONTRIBUTIONS BY

JONATHAN BEN-DOV, JACCO DIELEMAN, JUSTIN DOMBROWSKI, ERIK HERMANS,

AND MATHIEU OSSENDRIJVER

INSTITUTE FOR THE STUDY OF THE ANCIENT WORLD
NEW YORK UNIVERSITY PRESS
2019

ISBN 978-1-47980-176-3

Library of Congress Cataloging-in-Publication Data

Names: Bagnall, Roger S., editor. | Jones, Alexander, 1960- editor.
Title: Mathematics, metrology, and model contracts : a codex from late
    antique business education (P. Math.) / edited by Roger S. Bagnall and
    Alexander Jones ; with the assistance of Katja Kosova and Zoe Misiewicz ;
    and with contributions by Jonathan Ben-Dov, Jacco Dieleman, Justin
    Dombrowski, Erik Hermans, and Mathieu Ossendrijver.
Description: New York : Institute for the Study of the Ancient World and
    New York University Press, 2019. | Includes bibliographical references
    and index. | Summary: "Mathematics, Metrology, and Model Contracts is a
    comprehensive edition and commentary of a late antique codex. The codex
    contains mathematical problems, metrological tables, and model
    contracts. Given the nature of the contents, the format, and quality of
    the Greek, the editors conclude that the codex most likely belonged to a
    student in a school devoted to training business agents and similar
    professionals. The editors present here the first full scholarly edition
    of the text, with complete discussions of the provenance, codicology,
    and philology of the surviving manuscript. They also provide extensive
    notes and illustrations for the mathematical problems and model
    contracts, as well as historical commentary on what this text reveals
    about late antique numeracy, literacy, education, and vocational
    training in what we would now see as business, law, and
    administration"— Provided by publisher.
Identifiers: LCCN 2019032446 | ISBN 9781479801763 (hardcover) | ISBN
    9781479801787 (ebook other) | ISBN 9781479801770 (ebook)
Subjects: LCSH: Manuscripts, Greek (Papyri) | Paleography, Greek. |
    Mathematics—Early works to 1800.
Classification: LCC PA3300.L6 M38 2019 | DDC 487/.3--dc23
LC record available at https://lccn.loc.gov/2019032446

Contents

PREFACE

This edition and study of twelve leaves from a fourth-century CE Greek papyrus codex containing mathematical problems, metrological texts, and model contracts began as a graduate seminar held at the Institute for the Study of the Ancient World in 2010. The editors and contributors listed on the title page were all participants, and produced preliminary transcriptions and translations of individual codex pages on the basis of photographs. The final edition has been thoroughly revised and expanded by Roger Bagnall and Alexander Jones; Katja Kosova and Zoë Misiewicz prepared the index.

Our edition is arranged in three major parts: (I) an introduction to the manuscript, its various varieties of text, and its relations to other comparable texts in Greco-Roman papyri and similar archaeologically recovered media, (II) the edition proper with critical apparatus (to indicate standard orthography and graphic features) and facing English translation, and (III) commentaries to the texts. An index to the Greek texts is provided at the end. Much of the material in *P.Math.* is likely to be of interest to readers unfamiliar with the technicalities and conventions of papyrology, and for such readers we recommend an introduction to the field such as Bagnall 2009a. The codicological terminology is discussed in detail in Turner 1977.

The editors wish in the first instance to thank the anonymous owner of the nine leaves constituting the major remains of the codex for inviting us to edit them, for providing photographs as well as the opportunity to consult the originals, and for responding to our questions about them. The Walters Art Museum, Baltimore, assisted in various ways, including giving us the space and facilities for examining the leaves, and we particularly acknowledge the help provided by William Noel (then Curator of Manuscripts and Rare Books) and Abigail Quandt (Head of Book and Paper Conservation). We are also indebted to the Cotsen Children's Library, Princeton University, and its curator, Andrea Immel, for permission to study and edit the three leaves that are in the Cotsen's collection, and for supplying photographs.

In the final stages of preparation of the manuscript, we have been able to incorporate information about three unpublished documents or groups of documents containing material similar to what is found in the codex published here. We are grateful to Ruth Duttenhöfer, Jean-Luc Fournet, Todd Hickey, and Julia Lougovaya for information about these texts.

I. Introduction

1. The Manuscript

The present edition comprises twelve leaves from a papyrus codex (hereafter *P.Math.*), datable from its contents as well as on paleographic grounds to the fourth century CE, and containing texts of three kinds: mathematical problems, metrological texts, and model documents. Nine of the leaves (designated by the letters A through I, with E and F constituting an intact bifolium) were purchased by the anonymous owner of the "Archimedes Palimpsest" in 2001; the other three (designated M through O), around the same time, by the late Lloyd Cotsen, who donated them to the Cotsen Children's Library, Princeton University (CTSN Q 87167). Previous to these acquisitions, all the leaves had been in the hands of the antiquities dealer Bruce Ferrini.

The first knowledge of *P.Math.* in the scholarly community dates from the early 1980s.[1] In 1982 the Vatican Library received photographs (of very poor quality) of the bifolium and of leaves of a Greek papyrus codex of *Exodus*.[2] In the same year, Ludwig Koenen received photographs (not traceable, but likely the same ones) of *P.Math.* and the *Exodus* codex as well as of the Coptic codex now known as *Codex Tchacos* (containing the *Letter of Peter to Philip*, the so-called *First Revelation of James*, the *Gospel of Judas*, and the *Book of Allogenes*), and in May, 1983 Koenen, David Noel Freedman, and Stephen Emmel were allowed a brief examination at a hotel in Geneva of these three codices and a fourth, a Coptic manuscript of the letters of Paul.[3] The manuscripts were in extremely brittle condition, wrapped in newspaper and stored in three oblong boxes. The only available report of this examination, by Emmel, provides no details con-

---

1. For a more circumstantial account of the modern history of the group of four codices to which the mathematical codex belongs, see Krosney 2006. Rodolphe Kasser's account of *Codex Tchacos*'s history in Kasser, Meyer, Wurst, and Gaudard 2007, 1–12 omits any mention of the other three codices.

2. Minutoli and Pintaudi 2011, 194–195 and plate 24 (photograph of E verso). We thank Rosario Pintaudi for providing us with a scan of a second photograph showing F recto.

3. Emmel 2015.

cerning *P.Math.* They were again seen briefly in March, 1984 in New York by one of the present editors (RSB) under conditions that prevented the recording of detailed observations.[4]

Following Frieda Tchacos Nussberger's purchase of the four codices in 2000, they were held for several months at the Beinecke Library, Yale University. A report by Robert Babcock, then the Beinecke's Curator of Early Books and Manuscripts, supplies the earliest description of *P.Math.*:[5]

> At least 17 substantially complete leaves (including one complete bifolium) of a mathematical text, dealing with geometry (the measuring of triangles and liquid volume, among other things), and hundreds of small fragments. There are numerous drawings, some mathematical and related to the text, others appear to be purely decorative (crosses). Extensive searching indicates that the text cannot be identified with any known extant mathematical treatise from antiquity. The script suggests a fourth or fifth century date. No trace of the original binding is present, but the bifolium has sewing holes that show that the book was originally stab sewn.... The large leaves of this manuscript were placed in the front of the volume of Pauline letters..., with which they have no relationship. They do not belong to that binding, as the leaves are larger than that binding.

It thus appears, not only that Ferrini, who purchased *P.Math.* from Tchacos Nussberger later that year, was responsible for its partition into separate lots of leaves, but that at least five leaves and many smaller fragments that still existed in 2000 are unaccounted for, perhaps sold by Ferrini to other collectors.[6]

Unverifiable reports, relying on pseudonymous informants interviewed long after the claimed discovery, allege that the four codices were found in a tomb in the Jebel Qarara.[7] Internal evidence in *P.Math.* supports a general provenance in the Oxyrhynchite nome: one of the model documents, h3, is an agreement between (fictitious) residents of Oxyrhynchos and of the 6th pagus, and features of the model documents, including year numbers, are characteristic of documents from the Oxyrhynchite nome (see especially the commentary to text h3). That the codices were found together is credible, given the circumstances under which they were preserved when seen in the early 1980s; and the roughly contemporary Chester Beatty Codex AC.1390 (see below, section 3) even has mathematical problems in Greek and Christian literature (part of the Gospel of John) in Coptic in a single codex. The claim that the site was a tomb inspires less confidence but is not inherently unreasonable. Though the inclusion of a manu-

---

4. Krosney 2006, 148–149.

5. We thank Robert Babcock and Herb Krosney for providing us with this part of the report, in slightly variant versions.

6. Babcock (personal communication) recalls that all the substantial fragments were together at the front of the Pauline manuscript, and that there was a shoebox-size plastic container half filled with numerous small ("thumbnail or half-dollar size") bits of all four codices.

7. Krosney 2006, 9–27. For scepticism see Nongbri 2018, 95–96.

script of more or less practical mathematics in a funerary ensemble might seem odd to present-day expectations, it can be paralleled with the findspots of other ancient mathematical papyri such as *P.Cair. cat.* 10758 (late 4th or 5th century, also known as the "Akhmim Mathematical Papyrus"), reportedly found in the necropolis of Panopolis, and the Middle Kingdom hieratic Moscow Mathematical Papyrus (Pushkin State Museum), reportedly from the necropolis of Dra' Abu el-Naga'.[8] It is also not too far-fetched to bring into comparison recent discoveries of elite tombs in China, dating from the Warring States period through the Han Dynasty, in which were deposited collections of manuscripts of highly disparate genres, including mathematics.[9]

2. The Leaves and Their Sequence

The larger portion of *P.Math.* in the collection of the owner of the Archimedes Palimpsest consists of nine pieces of papyrus, of which seven are single leaves (A–D and G–I), one intact bifolium (E+F), and one smaller fragment (X) which joins along the top of I and, for the purposes of this edition, is treated as part of I. The portion in the Cotsen Children's Library consists of three single leaves (M–O). We estimate the original sheet dimensions as about 32 cm wide by approaching 30 cm. This would place the codex in Turner's Group 6, with 16 × 28 as the approximate leaf dimensions but with several representatives between 29 and 30 cm in height.[10] Peripheral damage to the leaves is most severe at the top: no significant upper margins survive, and many pages lack the top line of writing. Surviving binding holes show that the sheets were stab-bound.

Each recto or verso contains one or more texts, usually separated by decorative borders or by the diagrams that come at the end of many mathematical problems.[11] We have assigned each text a reference name consisting of the letter of the relevant leaf, in lower case, followed by the ordinal number of the text starting from the top of the (known or presumed) recto side. In only two instances (a5 and d4) does a text continue from the verso of one leaf to the recto of another; only once (b3) does a text run over from recto to verso; and in each case the overflow is small, as if the writer had miscalculated the line spacing needed to fit the text on one page. On other pages, when a text ends well above the bottom, the remaining space is partly filled with drawings of ankhs and other decorations. An index of the texts is provided at the end of this introduction.

In the following descriptions of the individual leaves, measurements are in centimeters, to a precision of half a centimeter except for the dimensions of the leaves themselves. The identifications of rectos and versos will be explained later in this section.

---

8. Baillet 1892, 2 (we reject Baillet's dating of the codex to the 6th century); Struve and Turajeff 1930, vii.

9. For example Zhiangjiashan tomb no. 247, Jiangling County, Hubei, excavated in 1983–1984, apparently belonged to a low-level official who was buried with manuscripts containing medical, legal, philosophical, and mathematical texts; see Morgan and Chemla 2018, 153–154.

10. Turner 1977, 18.

11. Such borders also appear, along with separators, in BL Add. MS 33369 (information from Todd Hickey).

A. 14.9 × 25.2. Margins: recto (↓) bottom 4.5, containing decorations, inner (2.0); verso (→) bottom 3.5, inner 1.5. Kollesis 5.5 from verso left edge. Remains of a binding hole 2.5 from bottom.

B. 14.4 × 26.2. Margins: recto (↓) bottom 2.5, inner 1.5; verso (→) bottom 4.0, inner 1.0. Top line of text on recto partially preserved. Kollesis 9.5 from verso left edge. Remains of a binding hole 2.5 from bottom, and binding hole 5 from bottom.

C. 15.2 × 27.1. Margins: recto (↓) bottom 2.0, inner 2.0; verso (→) bottom 1.0, inner 1.5. Apparent top line of text on both recto and verso partially preserved. Kollesis 13.5 from verso left edge.

D. 15.0 × 26.8. Margins: recto (↓) bottom 5.5, inner 1.5; verso (→) bottom 2.5, inner 0.5. No kollesis on verso.

E+F. 31.2 × 26.7. Margins: E recto (↓) bottom 1.5, inner 1.0 to binding fold, outer 0.5; E verso (→) bottom 1.0, no inner (writing runs across binding fold), outer 0.5; F recto (→) bottom 1.0, inner 1.0 to binding fold, outer 1.0; F verso (↓) bottom 1.0, inner 0.5 to binding fold, outer 1.0. Top lines of E recto and verso partially preserved. Binding holes on both leaves 2.0 and 4.5 from bottom. Kolleseis 13.5 and 31 from E verso left edge.

G. 13.3 × 26.5. Margins: recto (→) bottom negligible, inner 1.5; verso (↓) bottom 3.5, containing decorations, inner 1.5. Top line of recto and apparent top line of verso partially preserved. Kollesis 12.5 from recto left edge. Binding holes 2.0 and 4.5 from bottom.

H. 11.3 × 25.7. Margins: recto (→) bottom 3.5; verso (↓) bottom 10.0, containing decorations. Top lines of both recto and verso partially preserved. Kollesis 1.5 from recto left edge.

I (incorporating **X**). 10.0 × 26.7. Margins: recto (→) bottom 1.0; verso (↓) bottom 3.0. No kollesis on recto.

M. 16.2 × 25.9. Margins: recto (→) bottom 8.0, inner 1.5; verso (↓) bottom 6.5, containing decorations, inner 1.0. Kollesis 8–9 from recto left edge. Binding holes 2.5 and 5 from bottom.

N. 14.8 × 25.1. Margins: recto (→) bottom 5.0, outer negligible; verso (↓) bottom 1.5, outer 1.5. Kollesis 2.5 from recto right edge.

O. 16.0 × 26.9. Margins: recto (→) bottom 1.5, inner 1.5; verso (↓) bottom 1.5, inner 1.5. Top line of recto partially preserved. Kollesis 1 from recto left edge. Binding holes 2.5 and 5.0 from bottom.

We are not aware of any documentation of the order in which the extant leaves were arranged when they were discovered. Moreover, any page numbers that may once have existed were lost with the top margins.

As a working hypothesis, we suppose that the codex when complete comprised a single quire having the structure typical for fourth-century single-quire papyrus codices, namely that all sheets had their vertical-fiber sides outwards. [12] We know that in 2000 there still existed at least 17 leaves, so that the quire would have had at least 9 sheets. The extant bifolium E+F was the middle sheet, since text lines on E↓ (which is obviously E verso) run across the binding fold. Continuity of text from A→ to B↓ and from D→ to E↓ shows that these pairs of pages were consecutive, with their vertical-fiber sides as rectos. Binding holes (in some cases incompletely preserved as indentations along the broken edge of the leaf) establish or confirm that A↓, B↓, G→, M→, and O→ are rectos.

Continuity of horizontal fibers can be seen from C→ to G→, from G→ to B→, from B→ to M→, and from M→ to A→. Hence C+G and B+M constitute bifolia (in the case of C+G there is even a bit of common edge extant), and these bifolia and the one to which A belonged were originally adjacent pieces of the papyrus roll from which they were cut, making it very probable that B and C were consecutive leaves. Hence we have the following provisional sequence accounting for the eight leaves A–G and M:

> First half of quire (↓ recto): *m (≥ 0) leaves,* A, B, C, *n (≥ 0) leaves,* D, E
> Second half of quire (→ recto): F, *n+1 leaves,* G, M, *m+1 leaves*

Confirmation of this ordering of leaves A through E comes from consideration of their broken outlines (Fig. 1). A and B are both missing a roughly rectangular area at the top left of their vertical-fiber sides, in addition to other similarities of damage. C has a similar but slightly smaller loss in the same corner, with a less regular outline, and again in D there is a still smaller "bite" there, while E has what appears to be a small vestige of the same damage. Although the evidence considered above allows for one or more leaves between C and D, none of the extant leaves has an outline that would naturally fit into the progression in that place.

---

12. Turner 1977, 64–65.

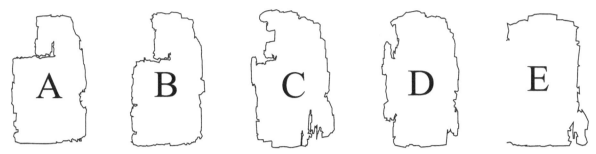

Fig. 1. Outlines of the vertical-fiber sides (rectos) of leaves known or presumed to have belonged to the front half of *P.Math*.

The bifolia E+F, C+G, and B+M preserve two kolleseis, separated respectively by 17.5, 14.5, and 13.5 cm, so that a rough average for the width of a kollema would be 15 cm. In Fig. 2, we reconstruct the portion of the original papyrus roll from which the five bifolia to which leaves A–G belonged were cut. Since from left to right the kolleseis show a gradual leftward trend against the pages, we can infer that a typical kollema was slightly less than half the breadth of a sheet, hence our estimate of 32 cm for the original sheet width.

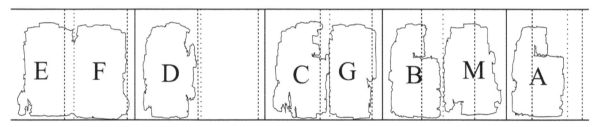

Fig. 2. Reconstruction of a portion of the original papyrus roll from which the codex bifolia were cut, showing the side with horizontal fibers. Extant or estimated kolleseis are indicated by broken lines, fold lines by dotted lines, and the boundaries between sheets by solid lines.

Turning to the remaining four leaves, it is probable that M, N, and O were originally together, since this would explain why these particular leaves were separated by Ferrini from the rest of the codex; moreover, the last mathematical problem on N↓ and the first on O→ (known to be O verso) are of exactly the same special type, namely determination of the capacity of a vaulted granary, and solved in the same incorrect manner, suggesting that these were consecutive pages. Hence we have the sequence of leaves F (missing leaf) G M (possible missing leaves) N O for the second half of the quire. H and I seem to offer no evidence for identifying which side was recto, but from the comparatively extreme damage to these leaves we may conjecture that they were either towards the beginning of the quire before A (preferably in the order I H) or towards the end after O (preferably H I). An argument favoring placing them at the front is that this would bring together the three model documents (our texts a1, h3, and i3). Fig. 3 shows a possible reconstruction of the part of the original roll from which the four leaves came, which would have been to the right of the part shown in Fig. 2. Another version would make N the conjugate of A and O the conjugate of either H or I.

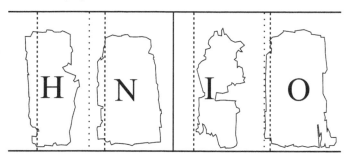

Fig. 3. Conjectural reconstruction of the rightward continuation of the roll from Fig. 2.

On the other hand, we have not been able to show fiber continuity between either N or O and any of A, H, or I, such as we might expect if H and I were from the quire's front half. (Since only a narrow strip of papyrus survives left of the kollesis on N and O, however, it is difficult to rule out any continuities categorically.) With H and I at the front, seven sheets of the presumed minimum nine would be partly or entirely accounted for by the extant twelve leaves (Fig. 4); with them at the back, eight leaves would be accounted for (Fig. 5). In the light of the uncertainties about where H and I were, our edition presents them at the end, following the ten leaves whose order is better grounded.

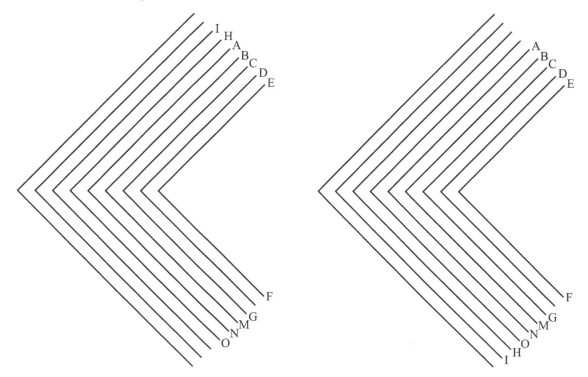

Fig. 4. A possible reconstruction of *P.Math.* as a nine-leaf quire with leaves H and I at the front.

Fig. 5. A possible reconstruction of *P.Math.* as a nine-leaf quire with leaves H and I at the back.

3. Dating

The model contracts provide several elements pointing to a date after the middle of the fourth century. The first is the use of the solidus in the loan of money (a1). Solidi are essentially absent from the documentary papyri before the Constantian monetary reform of 351–353, and as the commentary to a1 notes, we do not in fact have any published loans of solidi dated before 364. The use of myriads (restored, but inescapably) in i3.6–7 points in the same direction; myriads become a commonplace descriptor of sums of money only after the Constantian reform.

For this reason, the editor's date for the mathematical portions of Chester Beatty Codex AC. 1390, third–fourth century, with a preference for a date before 350, is almost certainly wrong. The Chester Beatty codex also has myriads of denarii. He also sought to use the supralinear curl as an indicator of amounts in the thousands as an indication of such a date (p. 35), but this is also a feature of our codex.[13] It is not an indicator of sufficient chronological precision to bear the weight that Brashear put on it. This codex seems likely to date from roughly the same period as our codex, i.e., the third quarter of the fourth century.

The second indicator is the use of a year 10 in h3.4–5. As the commentary there points out, of the two plausible dates in the fourth century, the earlier (342/3) would not be consistent with the monetary indications of the model contracts. The later would be the Oxyrhynchite era year 41=10, or 364/5. Mu (40) is also the most likely reading of the first digit of the year number in i3.

It is, of course, possible for the writer to have used Oxyrhynchite era years referring to a year earlier than that in which he was writing, but it would be unlikely that he would refer to any year later than the year following that in which he was writing; any more ambitious futuristic reference is excluded by the monetary changes, which could not have been foreseen, and the use of the era year itself, which would not antedate Julian's accession as Caesar in 355. Year 41=10 would in fact be only the second year in which the use of the regnal years of Constantius II and Julian can be properly seen as an era, following the death of Julian, so that neither year any longer refers to the reign of a living emperor. It seems on balance likely that 364/5 can be seen as the "dramatic date" of the composition of the model contracts, even if actual composition could have been slightly later.

4. Paleography

The first question that arises in looking at the writing of the codex is whether all of it was written by the same hand. It is, of course, easy to look simply at A recto and see that a1 and a2 reflect different styles of writing. The style of a1, like that of h3 and i3, is more rapid or "cursive" than that of a2 and the other mathematical and metrological portions of the codex. But neither does it look to us quite typical of the documentary hands found in contracts or official documents

---

13. Brashear, Funk, Robinson, and Smith 1990, 35. The date is given in DCLP, and thus Trismegistos, as 275–350.

of the fourth century written by professional bureaucrats or business agents. It is easy to get a synoptic view of these by looking at texts of a span of years in PapPal (pappal.info). Rather, the impression it conveys is of an attempt to write the kind of handwriting that we find in the mathematical texts, only in a more cursive fashion. That is a subjective judgment, to be sure, and the model contracts do at any rate resemble contemporary documentary handwriting far more than any literary hands. We can see no reason not to ascribe all three model contracts to the same writer.

It is not difficult to see considerable variation in the handwriting of the mathematical parts of the codex. Even inside a single page (one may take Or as an example), multiple forms of the same letter may be found (beta or epsilon, for example). But it is our impression that these variations do not exceed what may be found in the writing of a single individual, particularly over a period of time. For this reason, we take the codex to be a paleographical unit, on the view that an experienced writer could also have written in a similar but more cursive style appropriate to the different content of the model contracts.

Characterizing the hand of the mathematical sections is not simple. It is certainly not a professional book hand, although its general rightward slant is commonly found in literary texts as much as in documentary. The general character may be paralleled in some documentary texts; of those that appear on the PapPal gallery for this period there are similarities with *P.Oxy.* 14.1716 (333), *P.Oxy.* 66.4534 (335), *P.Oxy.* 63.4370 (354), *P.Oxy.* 48.3393 (365), and *P.NYU* 1.24 (373). But our writer, in the mathematical sections, is less professionally cursive than these. His script also bears some resemblance to teachers' models of the sort described by Raffaella Cribiore.[14] Her catalogue numbers 296 (Pl. XXXVI) and 389 (Pl. LXXIII) are the closest parallels of which she has illustrations. These are naturally without any indication of date; they have been assigned to the third to fourth centuries by editors, although Cribiore suggests that no. 389 may be later (a judgment with which we agree). It should also be said that the handwriting of the Chester Beatty Codex AC. 1390 mentioned earlier is, although more cursive than that of our codex, similar in many letter shapes.

5. Language

The Greek of the codex is far from flawless. The majority of the deviations from the norms of koine Greek are phonetic interchanges, mostly of a character well known from the documentary papyri of the Roman and Byzantine periods. These are listed below, with references in square brackets to the place in Gignac's *Grammar* (Gignac 1976–1981) where they are treated. Pure omissions of letters, signaled in the text with angle brackets, are not included. Similarly, but mainly in the mathematical problems, we find a number of morphologically non-standard forms. There are also some more complicated errors, which provide useful clues to the process of the creation of the codex. Because the model contracts present a somewhat different picture

14. Cribiore 1996.

from the metrological and mathematical texts, the phonetic and morphological phenomena in them are presented separately. The model contracts have an error rate of 25 percent (about 40 of 160 forms are nonstandard), almost all of them phonetic interchanges except for a few case or number errors, all in I verso. These phonetic errors are virtually all well paralleled in documentary papyri and require no special comment. It is common for the same word to be spelled both correctly and incorrectly on the same page.

But not all errors are plausibly explained as phonetic interchanges, as a look through the list below will indicate. It is more difficult to see what process could have produced them. The numerous other errors, including failure to use the correct case endings, are also readily paralleled in papyri written in this period. Such errors are generally explained as representing the difficulty that an Egyptian, whose native language did not decline nouns and adjectives, would have had in fully grasping the Greek system of accidence. Some other errors, entirely in the mathematical problems, seem to reflect mistakes in understanding the substance of the problems. These need to be considered in the context of the substantive errors of procedure made by the writer (or his source) in many of these problems.

PHONOLOGY

Model contracts.
    Vowels.
    ει > ι (G. 1.189–90): H ↓12, I ↓11
    ι > ει (G. 1.190–1): H ↓4, H ↓12, H ↓15
    ει > ε (G. 1.257–9): H ↓5
    οι > υ (G. 1.197–8): Ar3, Ar7, Ar10, Mr3
    οι > ι (G. 1.272): H ↓12
    η > υ (G. 1.264): H ↓12, H ↓15, I ↓9
    αι > ε (G. 1.192–3): Ar4, Ar8 (2x), H ↓10
    ε > αι (G. 1.193): Ar11, H ↓3, I ↓10, I ↓11
    ο > ω (G. 1.277): Ar11 (2x), H ↓7, H ↓9, H ↓14
    ω > ο (G. 1.276–7): Ar3, Ar7, Ar8, Ar12
    ο > α (G. 1.287–8): Ar3
    ου > ω (G. 1.208): H ↓3, H ↓6, I ↓8
    ο > ου (G. 1.212–13): I ↓8

    Consonants.
    δ > τ (G. 1.81): Ar4
    τ > δ (G. 1.80–1): Ar4, H ↓7
    κ > γ (G. 1.77): Ar5
    ρρ > ρ (G. 1.156): H ↓6
    σσ > σ (G. 1.158–9): H ↓7

Metrology and mathematics.

   Vowels.

      α > ο (G.1.286–7): Cr19

      ε > η (G. 1.244–5): Dr13, Dr19, Er13, Ov14

      ε > ι (G.1.249–51): Mv1

      ε > ο (G. 1.290–92): Fr2, Fr2, Fr9

      ε > αι (G. 1.193): Cr14, Mr13, Nr2

      η > ι (G. 1.235–7): Av11, Br5, Cv1 (also wrong gender)

      η > υ (G. 1.264): Bv11, Cr14, Cr19, Ev17, Mr11, Mr12, Or6, Ov7

      η > ω (not phonetic): Mr17

      η > ει (G 1.239–40): Br21

      η > οι (G. 1.266): Br11, Fr23, Mr6, Mv11, Nv8

      ι > α (not phonetic): Dr15, Dr19

      ι > η (G. 1.237–9): Bv12, Cv3, Ev10, Ev12, Ev16, Nr4

      ι > υ (G. 1.269–71): Dr4, Ev21

      ι > ει (G. 1.190–1): Dv5, Er11, Fr2, Gr3, Gr4, Gr6, Gr13, Gr19, I →8, Nr15, Ov7

      ι > οι (G. 1.272): I ↓13

      ο > ω (G. 1.277): Cr1, Cr20, Cv15, Cv16, Dr4, Dr18, Dv3, Dv24 (2x), Er5, Er6, Ev14 (2x), Fr3, Fv3, Fv18, Fv21, Gr8, Mr10 (3x), Mr16–17, Mv4, Mv6, Mv7, Mv9, Mv10, Mv11–12, Nr7, Nv2, Or24, Ov4

      ο > ου (G. 1.212–13): Dv19, Er2, Er11, Ev19, Ev22, Gr24

      υ > α (not phonetic): Cv13

      υ > ε (G. 1.273–4): Dv3, Nr5

      υ > η (G. 1.262–3): Ar17, Av2, Av4, Av6, Av13, Bv10 (2x), Bv11, Bv16, Cr15, Er5, Ev3, Ev4, Ev17, Gr2, Gr3, Gv5, Mr13, Mv4, Nv13

      υ > ο (G. 1.293): Br10

      υ > ω (G. 1.294): I →2

      υ > ει (G.1.272): Ev7, Ev16, Gr15, Gr16, Gr18, Gr19, Mr12

      υ > οι (G. 1.198–9.): Cr17, Cr10, Dv21, Er6, Er14, Fr1, Fv14, Gr6, Gr16, I →20, Mr15 (2x), Mr18, Mv7, Mv8, Mv10, Mv11, Mv12, Or11, Or24

      υ > ου (G. 1.215): Gr6 (by correction)

      ω > ο (G. 1.276–7): Br11, Cv6, Cv14, Dr18 (with ω then also written), Er12, Ev3, Fr16, Fv21, Gr1, Gr7, I →12, Nv2, Or3, Or16, Ov17

      ω > ου (G. 1.209–11): Ov3

      αι > ε (G. 1.192–3): Av14, Av17, Dr18, Er6, Ev8, Ev9, Ev10, Fr1, Fr19, Gr3, Gr9, Gr12, Gr14 (2x), Gr15, H →2, H →4, H →6, H →8, H →10, H →11, H →13, H →14, H →18, Mr5, Mv7, Mv8, Mv9, Mv11, Mv12, Ov17

      αι > η (G. 1.244–5): Gr10, Gr11, Gr16 (2x), Gr18, Gr19, H →15, I →1

      αι > ι (G. 1.249): Gr10

αυ > α (G. 1.226–8): Er14

αυ > αοω (cf. G. 1.230–31): Mr9

ει > ε (G. 1.257–9): Fr7, Fr8, Fr9, Or3

ει > η (G. 1.240–42): Er18, Mr12

ει > ι (G. 1.189–90): Av7, Bv14, Cr10, Cr15, Cr16 (2x), Cr17 (2x), Cr20, Dv4, Dv5, Dv9, Dv14, Dv15, Er17, Ev17, Fr1, Fr2, Fr7, Fr10, Fv11, Fv12, Fv15, Fv17, Fv20, Gr6, Gr9, Gr14, Gr16, Gr24, Gr24–25, Mr12, Nv5, Nv12, Nv18, Or6, Ov7

ει > ο (not phonetic): Gv17 (?)

οι > ε (G. 1.274–5): Cv1, Dr3, Mv2

οι > η (G. 1.265–6): Av5, Av13, Bv14, Gv9, Ov14 (?)

οι > ι (G. 1.272): Dr12

οι > ο (G. 1.200–01): Ev18, Gv5

οι > υ (G. 1.197–8): Cr10, Gv2, Mv5

οι > ου (G. 1.215: rare): Bv15

ου > οευ (G. 1.208): I →3

ου > ο (G. 1.211–12): Ar20, Br14, Bv14, Dr18, I →8 (2x), I →9

ου > ω (G. 1.208): Ar15, Av15, Fr14

Consonants.

γ > κ (G. 1.78): Dv3, Ev23, Gr12, Mr13 (2x), Or22

δ > τ (G. 1.81): Av1, Av4, Av6, Br12, Bv9 (2x), Bv14, Cr15, Cr19, Dr18 (2x), Er12, Ev9, Ev11, Fr2, Fr3, Fr8 (2x), Fr14, Gr12, H →11, H →15, Mr6, Mr11, Mr15, Mv8, Nv10

ζ > σ (G. 1.123): Nr7 (?)

ζ > σδ : Bv8

κ > γ (G. 1.77): Dr18, Mv2

σ > δ : Cr19, Nv2, Nv13, Nv15, Nv17

τ > δ (G. 1.80–1): Av 4, Av19 (2x), Cr15, Dr13, Dr19, Ev9, I →5, I ↓13, Nr15

γγ > γ (G. 1.116): Gr5

γγ > κ (G. 1.116): Dv19

γγ > νκ (G. 170–71): Er2, Er11, Gr22, Gv10

γκ > γ (G. 116): I →8, I →9

γκ > νκ (G. 1.168–9): Gr9, I →8

μβ > νβ (G. 1.168–9): Av19, Av20, Br8, Bv4, Cv6, Dv9, Dv11, Ev7, Ev19, Ev21, Ev22, Ev24, Gv6, Gv8, Mr7, Or19

μπ > νπ (G. 1.168–9): Mv12

ρρ > ρ (G. 1.156): Ov4

σπ > ψ (G. 1.154 on inversion; no examples of this type): Gr2, Gr11, H →1, H →4, H →8, H →11, H →12, H →13, H →18 σσ > σ (G. 1.158–9): I ↓13, Nv9

φ > π (G. 1.93): Er13, I ↓18

Omission of nu before consonant (G1.116–7): Br18, Dr4

Omission of final nu (G1.111–12): Dr6

Omission of final sigma (G. 1.124–6): Ev23, Fv7, I ↓18

Omission of word termination: Fr11, Or22

Insertion of nasal before stop (G. 1.118: normal in future of λαμβάνω and compounds): Mv8, Mv9, Mv11, Mv12, Ov15

Mistaken word: Fr11 (2x), Fr12

MORPHOLOGY (ALL TEXTS)

Masculine formation in place of neuter: Gr6

Masculine formation in place of feminine: Or2–3

Incorrect declension: Fv3

α instead of η as dative participial ending: Dv15

α instead of η as vowel in definite article: Or24, Or25

Formation on r-stem, ending in –ας: I →11

Third-declension accusative in –αν (G. 2.45–46): Cr19, Ov6

Uncontracted form (βορέας for βορρᾶς): Ov4

Thematic forms of –μι verb (προστίθομεν, συντίθω, συντίθομεν): *passim* (see index)

By-formation of present from aorist stem (ἕλω compound ὑφέλ- from root εἷλον, aorist of αἱρέω):

Av13	δὴ ἀπὸ τῆς κηνῆς βάσεως ἐφήλω τὴν κορη-
Av14	φὴν, ἀπὸ τῶν λ̄ οἰφέλωμεν ⲋ̄. λοιπὲ κ̄δ̄.
Dr6	[γί(νεται)] ῡ. ἀπὸ τῶ ῡ ὑφέλομεν ⲋ̄ν̄ⲋ̄. λοιπαὶ ρ̄μ̄[δ̄.]
Dr17	γί(νεται) ρ̄η̄. ἀπὸ τῶν [ῑ]ε̄ ὑφέλομαι θ̄. λοιπαὶ ⲋ̄.
Nv5	[ἐφ' ἑα]υτά. γί(νεται) Ἁ̄σ̄ρ̄ⲋ̄. τούτων ἐφελω τ̣[ὸ] δ̄. λο[ιπαὶ Ⲁ̄λ̄ο̄β̄.]
Er5	τά. γ[ί(νεται)] ⲋ̄ν̄ⲋ̄. τούτων ἠφέλωμεν τὸ τέταρτ[ον,]
Er6	ξ̄δ̄. ἀ̣[π̣]ὸ̣ τῶν ⲋ̄ν̄ⲋ̄ οἰφέλωμεν ξ̄δ̄. λοιπὲ [ρ̄ʼο̄β̄ʼ.]
Fv14	γί(νεται) ξ̄δ̄. ἀπὸ τῶν σ̄π̄θ̄ οἰφέλομεν ξ̄δ̄. λοι(πὰ) σκε. ὧν
Mv4	[ιε. ἐπὶ] τὸν η̄. γί(νεται) ρ̄κ̄. ἀπὸ τῶν τ̄ ἠφέλω'σʼμεν ‾ρ̄κ̄.
Or11	παρὰ τὸν δ̄. ⟨γί(νεται)⟩ ῑⲋ̄. ἀπὸ τῶν ῑⲋ̄ οἰφέλομεν δ̄. [γί(νεται) ῑβ̄.]
Or24	τὰν μίαν. γί(νεται) π̄ᾱ. ὧν πλευρὰ θ̄. οἰφέλωμ[εν]
Gv5	[ἀπὸ τῶ]ν υ ἠφέλομεν ⲋ̄ν̄ⲋ̄. λοπαὶ ρ̄μ̄δ̄. ὧν πλευρὰ ῑβ̄.

For the formation of thematic forms from τίθημι and its compounds, see Gignac 1981, 380–381 and Mandilaras 1973, 86. Active forms are much rarer than middle, and in documentary papyri participles are more common than finite forms. Forms of ὑφαιρέω occur in Chester Beatty AC.1390, but they are abbreviated after ὑφελ, leaving the mood uncertain. In this respect it is very similar to the Akhmim codex, where again the form is abbreviated. Brashear, like Baillet, resolved these as aorist imperatives, which is plausible in light of the use of imperatives in these manuscripts, a respect in which they differ from *P.Math.*, where the first person is consistently used (sometimes singular, sometimes plural). In the Akhmim codex, in some places there is no indication of whether indicative or imperative would be more correct. In *PSI* 3.186v, as reedited in Shelton 1981b, υφελω occurs. Shelton took this as a future and accented accordingly. But it is in light of *P.Math.* more likely that this is in fact a present indicative. The absence of such presents outside mathematical texts may lead one to see these forms as instances of the confusion of tenses (on which see Mandilaras 1973, 57) rather than a true birth of an alternative present.

Syntax

Model contracts.
Accusative for nominative: Ev19, Ev22
Genitive for dative: H ↓11
Accusative for dative: H ↓10
Plural for singular: H ↓10
Singular for plural: H ↓9

Metrological and mathematical texts.
Nominative for genitive: Cv4, Er4, Fv12, Mr7
Nominative for accusative: Br7, Bv3, Dv14, Fv2
Nominative for dative: Dv15
Genitive for nominative: Av12
Genitive for accusative: Br6, Fv19, Fv22, Ov6
Dative for nominative: Gr15, Nv11, Or20
Accusative for nominative: Cr17, Dv12–13
Accusative for genitive: Mv5 (2x), Nv19, Ov13
Accusative for dative: Dr15, Dr19
Singular for plural: Fr7, Gr11, H →3, H →4, H →6, I →7, I →18
Plural for singular: Gr9, H →7, I →10, Ov13
Participle for finite verb: Mr4
Omission of augment: Cr16
Second person singular for third person plural: Cr16
Neuter for masculine: Mr15
Masculine for neuter: Dv22

6. The Model Documents

Three of the texts in the codex, of which only one is fully preserved, fall into this category, a1, h3, and i3. What do we mean with the term "model contract"? Model contracts are to be distinguished from actually executed contracts, which should have contained official dating formulas (by the consuls, at the period to which we assign our codex) and the subscription of the acknowledging party or parties and of their *hypographeus* in the case of an illiterate party. Actual contracts also stand on separate sheets of papyrus and are not simply portions of pages of a codex. Because we have the endings of all three of the model contracts, we can be certain that they lack the concluding elements needed in actual contracts.

Somewhat more complex is the distinction between a model contract and a draft of a contract. An example of a draft contract, in this case a rider to a contract, has recently been published as *P.Bagnall* 47, with a detailed commentary by David Ratzan. Like our model contracts, this lacks the elements of execution, including the subscriptions, that would be needed in actual contracts, but unlike them it does not attempt to include the names of parties at the start, starting baldly with χαίρειν. It is thus pure draft legal language. In contrast, we have the names of the parties, even if imperfectly preserved, in h3, and it is clear that they also stood in a1.

The question of distinguishing actual from model contracts is not always simple, however, particularly with the presence of elements like the names of parties. This issue was confronted by Rosario Pintaudi and P. J. Sijpesteijn in the introduction to the volume of wooden and wax tablets in various collections (*T. Varie*, p. 8): "Alcuni dubbi si hanno a proposito di alcune tavolette lignee della Biblioteca Apostolica Vaticana (1, 2, 3, 8, 10, 11): si hanno testi di tipo documentario (contratti), scritti da scribi con una certa esperienza grafica, per quanto l'ortografia in molti casi risulti orribile; le parti coinvolte sono menzionate con l'indicazione delle località di provenienza, che spesso risultano attestate qui per la prima volta. Tutti elementi questi, che ci porterebbero a considerare i testi come veri e propri atti giuridicamenti validi. A questo si oppone il fatto che nessuno di questi contratti risulta però completo, anche quando lo spazio a disposizione nella tavoletta ne avrebbe permesso la conclusione regolare (formule finali, sottoscrizioni e sim.), senza presupporre il seguito su altre tavolette ora perdute! Non si può quindi escludere che anche quei testi più manifestamente documentari siano di provenienza scolastica, cioè scritti dal maestro perché gli studenti li potessero copiare ed acquisire confidenza con formule e situazioni quanto più vicine alla realtà."

When those words were written thirty years ago, the corpus of comparable texts was small (see *P.Rain.Unterricht*, where only no. 108 is really similar), but the analysis remains sound. With the benefit of the material provided by *P.Math.*, we can add that the combination of model contracts with texts of other types, including mathematical, in a codex appears as a diagnostic element for the scholastic use of the texts and thus their model character. This is particularly striking in the case of the tablets if one takes into account the probability that the tablets in that volume and others cited by the editors on pp. 7–8 belong to a single find, a large part of which was school texts of one sort or another, many of them mathematical, including tables and problems. Because that tablet find belongs to a much later period than *P.Math.* (the sixth and seventh centuries), the publication of *P.Math.* adds importantly to our knowledge of the use of model contracts in the school curriculum.[15] It is significant to note that only h3 concerns an area of land, a type of business transaction in which one might suppose a direct relationship with the geometric knowledge that the problems of *P.Math.* tried to convey.

The corpus of published model contracts will be enlarged in the near future by the publication of several other texts. We cannot incorporate their information fully into the present edition, but, thanks to their editors, we can briefly mention them and some of their chief characteristics. Most extensive is BL Add. MS 33369, a codex of ten wooden tablets from the Panopolite nome and to be dated to the fifth century. Each leaf is broken vertically, with a loss of about a quarter to a third of the width, but a substantial amount survives.[16] It has points of similarity with our codex, but also differences; see the commentary to a1 below for more detail.

Also on tablets, but apparently separate items rather than forming part of a codex, is a series of school texts in the Yale collection, which are now in press and will appear as *P.Yale* 4.184–191, edited by Ruth Duttenhöfer. These date to the seventh century and include a model receipt for *annona civica*, loans for repayment in kind, Coptic letter formula, and Greek contract formulas, along with multiplication and fraction tables.

7. The Metrological Texts

P.Math. contains five metrological texts: e3, g1, h1, i1, and m1. Like the model contracts, they all start at the top of a page, on the recto side except for e3. Texts e3 and m1 occupy an entire page; the rest are followed by mathematical problems. Perhaps this reflects how lessons were organized, with mathematical exercises taking up any time remaining after other material.

The first four texts all have extensive word-for-word parallels in other papyri, and parts of m1 have less exact parallels in e3:

> e3: *P.Oxy.* 4.669 (late 3rd or early 4th century).
> g1: *P.Oxy.* 4.669.
> h1: *P.Lond.* 5.1718 (late 6th century), *P.Ryl.* 2.64 (4th or 5th century).
> i1: *P.Oxy.* 1.9 + 49.3456 (3rd or 4th century),[17] *P.Oxy.* 49.3457 (1st or 2nd century), 3458 (3rd century), 3459 (3rd century), and 3460 (2nd or 3rd century).

---

15. We are indebted to Jean-Luc Fournet for the information that model contracts also occur in a codex of tablets that he is to publish together with Todd Hickey.

16. The catalogue record may be found at https://urldefense.proofpoint.com/v2/url?u=https-3A__tinyurl. com_yxrvt4ug&d=DwIFaQ&c=slrrB7dE8n7gBJbeOog-IQ&r=a1GzvfjAQ4FOT67EmNqRIAkSUOmrC2-_ N3pEK_8DkHo&m=TLnpq-YcNRech5Y7m-YFuehiSwy46bulYkF81aaLIAc&s=EixQj4BL2WDrK- _6oU1olOlC574WrB-jVWmVMHfP-Dc&e=, derived from M. Richard, *Inventaire des manuscrits grecs du British Museum I, Fonds Sloane, Additional, Egerton, Cottonian et Stowe,* Paris 1952, p. 56. It is to be published by Jean-Luc Fournet, Todd Hickey, Yasmine Amory, and Valérie Schram.

17. We are indebted to a referee for the information that a full edition of these tables will appear in the forthcoming Festschrift for G. Messeri, arguing for a third-century date.

Thus the teaching of systems of units involved the copying—and memorization?—of a few standard texts that circulated in some cases for several centuries, with some allowance for variation through reordering, omission, or insertion of material.

The first part of g1 (G recto 1–6) represents an initial approach to learning units, through lists of their names in order of magnitude, in this instance increasing and beginning with units of length followed by units of area. The main object of the metrological texts, however, was to teach how the various units were quantitatively related, so that one could perform unit conversions such as we encounter in many of the mathematical problems.

The relations expressed between units fall into two kinds, depending on whether the units in question measure the same kind or different kinds of magnitude. A relation between two units of the same kind, say two units of length, is expressed as how many of one unit are contained in one of the other. In the metrological texts in *P.Math.*, relations between units of different kinds are limited to definitions of units of area or volume in terms of units of length applied to the two or three dimensions; we do not find, for example, a definition of a unit of weight in terms of a specified volume of some material. A unit appropriate for a particular kind of object or commodity may, however, be related to a less particularized unit of the same general kind, as in e3 (E verso 16–18) and m1 (M recto 12–13) where equivalents are given of the solid cubit, a general unit of volume, in volume units suitable for grain (artabas) and for liquids (metretai).

The texts make much use of syllogism-like statements on the pattern, "Unit A contains *l* unit Bs, and unit B contains *m* unit Cs, so that unit A contains *n* unit Cs" (where $n = l \times m$). This formula may have been preferred for ease of memorization, since it connects three or more units of the same kind and provides an arithmetical relation that can be verified. Text i1 consists entirely of quasi-syllogistic statements, which do not appear to follow a larger-scale plan, since the kinds of unit in each statement have no obvious connection with the ones that precede or follow.

Text h1, which does not employ quasi-syllogistic statements, is the most systematically organized. It provides relations among eleven units of length, devoting to each, in order of diminishing magnitude, a sentence listing the equivalents in each of the smaller units, again in order of diminishing magnitude, so that by the end a direct relation has been given for each possible pair of units. (For the extraneous last sentence of the text, H →16–19, see the commentary.)

8. Relations in the Metrological Texts

This section summarizes the relations between units in the five metrological texts e3, g1, h1, i1, and m1. In the tables, the quantities in the cells represent the number of the unit named at the top of the column contained in the unit at the left end of the row. Square brackets and angle brackets have the same meaning as in the transcriptions, while italics indicate quantities that we have included as supplements for the user's convenience. When a table merges data from more than one text, quantities or digits of quantities that are preserved in at least one text are not bracketed. We have not incorporated data from other papyri. Since a large part of the metrological system is reducible to defined relations to the cubit, orders of magnitude in terms of modern units can be estimated from the approximate equivalent of one Egyptian cubit to 0.53 meters; see also Bagnall 2009b.

length (g1, h1, i1).

	hamma	reed	xylon	pace	cubit	foot	spithame	lichas	palm	finger
surveyor's schoinion	[8]	[16]	[32]	[48]	96	144	192	[288]	576	<2>304
hamma		2	4	6	12	[18]	24	36	72	[288]
reed			2	3	6	[9]	12	18	36	[144]
xylon				1½	3	4 [½]	6	9	18	[72]
pace					2	3	[4]	6	12	48
cubit						1½	2	3	[6]	24
foot							1⅓	[2]	4	16
spithame								1½	3	12
lichas									2	8
palm										4

length (e3)

	ogdoon	cubit
linear schoinion		100
surveyor's schoinion	[8]	96
ogdoon		12

length (e3)

	royal xylon	cubit	palm	finger
surveyor's schoinion	32			
royal xylon		3	18	[7]2

	private xylon	cubit	palm	finger
surveyor's schoinion	36			
private xylon		2 2/3	16	64

length (g1)

	Ptolemaic (Egyptian) foot	Italian foot	builder's foot
palm	4		
finger	16	13⅓	13⅔

length (g1)

	weaver's cubit (pygon)	builder's cubit (public cubit)	nilometric cubit	loom cubit	?
palm	5	6	7	8	9

length (m1)

	mile	gues	stade	surveyor's schoinion	cubit
schoinos	4	12	60	240	23,040
mile		3	15	60	5760
gues			5	20	1920
stade				4	384
surveyor's schoinion					96

area (m1)

	amphodon	laura	area cubit
dodekatikon	[90]	180	
amphodon		2	20,000
laura			10,000

area (e3)

building-site cubit = [100] area cubits

area (e3, m1)

	bikos	area cubits
country aroura	48	9216
urban aroura (aroura for building sites)	50	10,000
(country) bikos		192 (text: 992)
(urban) bikos		200

volume (e3, m1)

1 (public) naubion = 1 xylon × 1 xylon × 1 xylon = 27 solid cubits

1 private naubion = 18 + 1/2 + <1/3> + 1/9 + 1/54 (i.e. 18 26/27) solid cubits
(i.e. 1 private naubion = 1 private xylon × 1 private xylon × 1 private xylon)

1 solid cubit = 3 + 1/4 + 1/8 artabas (i.e. 3 3/8) = 3 metretai
(i.e. 1 artaba = 1 foot × 1 foot × 1 foot)

volume, liquid capacity (i1)

	chous	kotyle
metretes	12	144
chous		[12]

volume? (h1)

	hamma	reed	xylon	pace	cubit	foot	spithame	lichas	palm	finger
naubion	[2¼)	4½	9	13½	27	[40½]	[5]4	81	[162]	648

(Note: this seems to be a non-standard scale of units, combining the familiar relation 1 naubion = 27 solid cubits with the relations of linear cubits to other length units.)

weight (i1)

	tetarte	thermos	carat
mnaeion	16	96	192
tetarte		[6]	[12]
thermos			2

weight (i1)

	uncia	semiuncia	gramma
litra	12	[24]	288
uncia		[2]	24
semiuncia			1[2]

weight/currency (i1)

	mina	stater	drachma	tetrobol	obol
talent	60	1500	[6000]	10,500	4[2,000]
mina		25	100	175	700
stater			[4]	7	28

9. The Mathematical Problems

Most Greek mathematical papyri contain problems, arithmetical tables such as multiplication tables and tables of fractions, metrological texts, or a combination of these genres. The normal format for presenting problems is as a series of statements and solutions, with no intervening commentary (see Appendix A). Successive problems are often thematically related (e.g., *MPER NS* 1.1 consists of problems concerning volumes of three-dimensional figures such as parallelepipeds and pyramids, and *P.Chic.* 3 of problems concerning areas of polygonal fields). 44 problems are preserved in *P.Math.*, most of them complete or with enough surviving so that we can recover what is to be solved, the method of solution, and the results. This is not quite the largest known collection of problems in a Greek papyrus—*P.Cair. cat.* 10758 has 50—but it is the most varied collection, and especially rich in geometrical problems. There are occasional instances of thematic grouping (e.g., d4, e1, e2); but for the most part each problem seems to have little relation to the preceding and following problems, and in the leaves A–G, whose order is well established, we see little sign of a progression from more elementary to more advanced problems.

In common with the problem texts in other Greek and Demotic mathematical papyri as well as in those from many ancient and medieval societies, most problems of *P.Math.* are about how to find unknown quantities in scenarios defined at the outset in terms of given quantities.[18] The most common pattern begins with the specification, as a nominal phrase, of an object un-

18. Beyond the broad characterization of the problem texts delineated above, the Greek and Demotic problem texts have surprisingly little in common, either with respect to their scenarios or the algorithms employed to solve them, the exceptions including a few that were practically universal to the genre across cultures, such as the Diagonal Rule ("Theorem of Pythagoras"), Quadrilateral Area Algorithm ("Surveyor's Formula"), and algorithms for a circle's area that assume the approximation π = 3. For the corpus of Demotic mathematical papyri see Parker 1972 and 1975; note that in the former volume Parker followed the unusual practice of numbering the mathematical problems continuously through all the papyri.

der consideration, which may be simply a "real-world" entity such as "a breach in a dike" (a2), or such an entity with an added indication of its shape, such as "a circular excavation" (d4), or a mathematical abstraction such as "an isosceles triangle" (c3). Then a participial or relative construction may introduce the given dimensions or other quantities, usually expressed as a number of units. (Sometimes the units are not named at this point but become explicit later in the text.)

The solution, optionally introduced by a phrase such as οὕτως ποιοῦμαι ("I proceed as follows"), consists of a series of arithmetical operations and their results, culminating in a result that is the solution of the problem. The writer of *P.Math.* is fond of concluding a problem text with οὕτως ἔχει or οὕτως ἔχει ὁμοίως, a phrase that we interpret as an assurance that the same procedure is applicable to similar problems with different data ("This way for similar cases," cf. the alternative phrase καὶ ἐπὶ τῶν ὁμοίων, "And (so) for similar cases," at the conclusion of n3, N verso 6).[19]

The operations may be expressed using first-person (singular or plural) indicative verbs or by verbless prepositions (e.g., ἐπί for multiplication, παρά for division); imperatives are not used in *P.Math.*, though they are common in other mathematical papyri. While the arithmetical operations are typically not explained, exceptions are frequent, such as the following instance where the numerical operation is preceded by both an explanation that this is a unit conversion and by the specific relation of units (d4, D verso 21–23):

> I convert the naubia to cubits. One naubion contains 27 cubits. 21 1/3 times 27.
> The result is 576.

Explanatory supplements are not limited to metrology, but can also express a step of a computational algorithm in generalized terms relating to the scenario before repeating the step using the numbers in the particular problem. In the following passage of a5 (A verso 12–15), we have not only this kind of functional explanation preceding an arithmetical operation, but also, following an intermediate result, a statement of what that result means in terms of the geometrical configuration:

> Since from the common base I subtract the top (dimension), we subtract 6 from 30. The remainder is 24. Half of this is 12. This will be the base of the right-angled (triangle).

Such intermittent explanations suggest that the teaching of algorithms in the classroom made more use of abstraction and generalization than was expected in the written solutions, and that

---

19. The same phrase, however, concludes the metrological texts e3 (E verso 24) and h1 (H →19, curtailed to merely οὕτως), where such an appeal to generalization makes no sense. Cf. also *P.Gen.* 3.124 i.9–10: ὁμοίως δὲ καὶ ἐπ᾽ ἄλλων ἀριθμῶν εὑρήσομεν, "We will find similarly for other numbers."

the student was expected to have some comprehension of the meaning (in terms of the scenario) of intermediate results.[20]

Alternatively, an operation may be followed by a question and explanation, as if a teacher is responding to a student's oral question (or conversely, as if the teacher is quizzing the student), as in the following example (n1, N recto 6–8):

> The result is 77,760. I divide this by 27. Why? (Because) one naubion [contains]
> 27 cubits. The result is 2880.

Or the teacher may, as it were, ask the student an arithmetical question (f2, F recto 10–13):

> 200 contains how many hundreds? 2.
> Three hundred contains how many hundreds? [3.]
> 400 contains how many hundreds? 4.
> I add 2 and 3 and [4.] The result is 9.

A few of the problem texts in *P.Math.* (f2, m2, and perhaps d3) conclude with a verification that the result obtained at the end correctly satisfies the conditions of the problem.

The solutions in problem texts in Greek mathematical papyri such as *P.Math.* have the appearance of being models possessed by the teacher in advance of presenting the problems to students. Even the question-and-answer passages in *P.Math.* were likely scripted, not genuine transcripts of interchanges that occurred in the classroom; they can be paralleled in other mathematical papyri.[21] The most crucial decisions that the solver of a problem must make were never written down. The student was expected to know a repertoire of algorithms applicable to a variety of general scenarios, such as algorithms for finding the area of a rectangle or of any quadrilateral figure given the sides (see next section); so the first questions to be addressed in the face of a new problem were, which algorithm applies, and how are the given data to be matched up with the givens assumed in the algorithm? In some cases, a complete solution may have called for one algorithm applied to part of the data, and another applied to the result of that algorithm together with the rest of the data. Algorithms sometimes also had to be transformed so that what was originally the result becomes a given while one of the original givens

20. See also *P.Lond.* 5.1718, 71–77 for statements of geometrical problems, first in generalized terms and then with specific numbers.

21. *P.Cair. cat.* 10758 has numerous examples of the interrogatory formulas ἐν ποίᾳ ψήφῳ (asking what division of whole numbers would result in a given expression of unit-fractions), τί ἐπὶ τί (asking for whole-number factorizations of a given number), and ἀπὸ N πόσα (asking what a given fraction of N amounts to). Chester Beatty Codex AC. 1390 also has ἐν ποίῳ ψήφῳ and τί ἐπὶ τί questions (1.24–25, 3.4, 3.20–21). In the same manuscript are several διὰ τί ("why?") questions concerning metrological conversions (1.10–11, 2.25–27, 3.11–12, cf. 2.23 for a metrological question not using the διὰ τί formula) and an arithmetical operation (3.28–4.1). Another metrological διὰ τί question is in *PSI* 3.186.6–8, for which see Shelton 1981b, 100, but we propose restoring thus: [γί(νεται) 'Αυμ. 'Αυμ ἐπὶ] τὸν γδη. διὰ τί ἐπὶ τὸν | [γδη; ὅτι ὁ σ]τερεὸς πῆχ(υς) χωρήσι ξηροῦ | [ἀ(ρτάβας) γδη. ἄρα ὁ θησαυρὸς (?)] χωρήσι ξηροῦ ἀ(ρτάβας) 'Δωξ. Shelton, p. 100, cites an instance of a διὰ τί (concerning an arithmetical operation) in pseudo-Heron, *Stereometrica*, ed. Heiberg p. 140.2.

becomes the result; for example an algorithm for finding the area of a plane figure from its linear dimensions may be turned into an algorithm for finding one of the linear dimensions from the others together with a known area. The artificiality of such inverted problems shows that the goal of problem-solving was not just preparedness for routine and repetitive practical situations but a certain degree of mathematical versatility. The calculations involved in the arithmetical operations, even when involving large numbers and fractions, are omitted in the texts, and it is not clear whether these were actually performed in the didactic setting or whether the results were taken for granted.

Taken as a whole, the mathematical problems in *P.Math.* are riddled with mistakes. These are of diverse kinds, and offer clues about the processes underlying the composition of the manuscript and hence the way that mathematical skills were taught in fourth-century Egypt. First of all, there are problems that presume mathematically impossible scenarios. In b3, a rectangular wall of constant thickness is presupposed, such that the outer and inner perimeters and the thickness are treated as independent data, whereas in reality once one has chosen any two of these, the third is determined—and the particular given quantities in b3 are grossly inconsistent. Problem o2 is of the artificial, inverted algorithm kind, presupposing a quadrilateral with three sides and the area known and requiring the fourth side to be found. A figure with the prescribed dimensions, however, turns out to be impossible, and the only available algorithm that a student could apply, which is by its nature approximate, leads to the absurd result that the quadrilateral degenerates to a straight line with zero area. These are blunders that one would hope would have been exposed as soon as the solutions were attempted in the classroom.

Next, we have instances in which the problem-solver has chosen an algorithm that is inappropriate for the problem's scenario, indeed for any plausible scenario. Problem b4 is set out as a very simple demand to find the area of a rectangular field with given dimensions ("side" and "base"). Instead of just multiplying the two numbers, the solution begins by squaring the "side," and then multiplies the product by the "base," as if the object was to find the volume of a prism with a square cross-section. The solution is rescued by a final division by the "side," but one can only wonder what triggered this diversion into three dimensions. Still stranger are two forays into the *fourth* dimension, n4 and o1, in which the scenario asks for the volume of what is apparently a box-like shape with a vaulted roof, so that dimensions are given for the length, width, and depth of the box part and (we suppose) for the height of the roof. The intended algorithm would perhaps have consisted of calculating the volume of the box, and separately the volume of a rooflike figure having the given vault height and the rectangle formed by the given length and width as base. Instead, all four quantities are multiplied together, and the product is considered to be the volume.

The solver also occasionally introduced spurious steps in an algorithm, spoiling the result. In g2, for example, an intermediate calculation of a circle's area from its known circumference, a division of the circumference by 2 is performed that has no place in the algorithm, causing the problem's solution to be a quarter what it should be. In n2, where the algorithm requires division by 2, the solver divides instead by 4, so that the solution is too small by half.

If the solver's grasp of the algorithms was sometimes shaky, he nevertheless appears to have been well grounded in basic arithmetic; there seem to be no errors of calculation throughout the problems. Multiplications that involve fractions and that result in products on the order of hundreds of thousands are correctly executed (e.g., b3, i2, and n4). Divisions usually lead to whole-number results. Square roots also always turn out to be whole numbers, never exceeding 30, such as could have been found in a table of squares.

Lastly, we have copying errors. A few isolated missing or miscopied digits in the numerals (e.g. 32 for 12 in d2, D recto 13, and missing unit-fractions in o5, O verso 6 and 8) just show that we are reading a transcript. More revealing are passages in the solutions as they appear in *P.Math.* that are clearly distortions or misunderstandings of an underlying correct version. Thus in f5, the area of a field has been found as 1176 area cubits, and this is to be divided by 9216 to get the area in arourai. These numbers are correctly written in the manuscript, and the original solver of the problem certainly performed the division correctly, obtaining 1/8 + 1/384. In the manuscript, however, the first fraction has somehow become the symbol representing 4000, so that the figure 4384 is presented (twice!) as the number of arourai in the field, a result whose absurdity should have been obvious both from the relative sizes of the numbers involved in the division and from a common-sense realization that area of a field whose longest side is just 90 cubits can scarcely exceed one aroura, let alone several thousand. Again in f2, the passage quoted above with its three questions asking how many hundreds are contained in 200, 300, and 400 is turned into nonsense through repeated confusion of the word for "hundred" (ἑκατόν) with that for "each" (ἕκαστος) or for "hundredths" (ἑκατοστή). Such a series of errors are symptomatic of incomprehension, not mere inattention, and could hardly have been made by the original solver.

The diagrams in *P.Math.*, which were presumably drawn and certainly labelled by the same hand that wrote the texts, are among the manuscript's odder features. Diagrams were a common, though not indispensable, accompaniment to problem texts in mathematical papyri, usually drawn below, but sometimes preceding, the text. Unlike the diagrams of Greek deductive geometry, they were not integral to the logic of the texts, but served to visualize the scenario, data, and results in a highly schematic manner. For example, if a problem hypothesizes a quadrilateral field with given lengths of its north, south, east, and west sides and required one to find the area, the diagram could consist of a crudely drawn quadrilateral with the four sides labelled with the direction names and the numerical lengths, and with the numerical area inscribed inside the figure.

Problems e1 and e2, dealing with cylindrical pits, have diagrams that are reasonably normal for the genre. In e1, the givens are the "upper diameter," that is, the diameter of the circular top face of the cylinder (16 cubits), and the depth (3 cubits), and one is required to find the volume (21 1/3 naubia). The diagram consists of a crudely drawn circle with a diameter drawn across it from top to bottom; there is no attempt to represent the pit as a three-dimensional object. Written outside the circle are the numerals 16 (near the upper end of the diameter), 3 (near the diameter's lower end), and 21 1/3 (to the circle's right). There is no text other than these numbers, and only from the text can one know which number represents which element in the problem. In e2, we are given the pit's depth (9 cubits) and its volume (16 naubia), and we are asked to find the circumference (24 cubits). The diagram is just a crude circle with the numerals 24, 16, and 5 written outside it in roughly the 12 o'clock, 3 o'clock, and 6 o'clock positions. Thus the numbers for the depth and volume occupy the same position as in e1's diagram, so there is at least some consistency of layout.

Problem a5 has a more complicated geometrical scenario, a trapezoidal figure that can be partitioned into a central rectangle flanked by two similar right-angled triangles. The givens are the four sides, namely 6 schoinia for the "top" (i.e., the shorter parallel side), 30 schoinia for the "common base" (i.e., the longer parallel side), and 15 schoinia for the "legs" (i.e., the two sloping sides). In the solution part of the text, various intermediate results are obtained, including the base of the right-angled triangles (12 schoinia), the square of this base (144, units not specified but actually arourai), the square of the triangles' leg (225), the altitude of the trapezoid (9 schoinia, for some reason called the "base of the rectangle"), and the areas of the triangles and the rectangle (in each case 54). The expected final result, namely the total area (162 arourai), is omitted. The diagram shows the trapezoid partitioned by vertical lines into the two triangles and the rectangle, as expected (compare our own diagram in the commentary to this problem). Again there is no text, only numerals. Of the givens, the 15 for the "legs" is written next to the right sloping side, but also above the upper parallel side where we might have expected this side's length, 6. In fact neither 6 nor 30 appears anywhere. 12, the base of the triangles, is written in a central position below the rectangle, and 9, the altitude, is written just above the 12 inside the rectangle. The square numbers 144 and 225 are written near the sides of the right-side triangle to which they pertain, and 54 is written inside both triangles (but not inside the rectangle). All in all, the selection and placement of the numerical data in this diagram are so wayward that it would surely have been a source of confusion to anyone using it as an aid to following the text.

Elsewhere it is not merely the placement of the numbers in the diagrams that seems disconnected from the problems as written, but the very way that a geometrical scenario is imagined. In g4 the statement of the problem admittedly invites trouble. Though the beginning is mutilated, it is evident that the (three-dimensional) object in question is described as "circular" (στρονκύλουν, *scil.* στρογγύλον), which in other problem texts means a solid shape having a circular cross-section, namely a cylinder, cone, or conical frustrum; however, the givens are "length," "width," and "thickness," which would be appropriate for a parallelepiped. The solver who wrote the solution ignored the alleged circularity, and applied an algorithm appropriate for a parallelepiped. The diagram, however, consists of two horizontal parallel lines joining two small circles (the left one is incompletely preserved), evidently representing a cylinder—though grossly out of proportion with the rather thin slab implied by a figure whose thickness is in fingers while the other dimensions are in cubits.

In other cases, however, it is hard to imagine how the person who drew the diagram imagined that it represented the problem's scenario, even though we usually find at least some of the data from the problem scattered around the drawing. Thus c5, a problem about a rectangular parallelepiped, is accompanied by a drawing looking a bit like the reverse side of a letter enve-lope, such as might be meant for a triangular prism. On the same page, problem c3 concerns finding the area of a plot of land explicitly described as an isosceles triangle, but the diagram shows a more or less right-angled triangle with an additional line crossing it parallel to its shorter leg. Nor is this a one-time accident, since two other problems about isosceles triangles, d1 and g3, have similar drawings. Problem o2 concerns a quadrilateral, yet its diagram consists of a single horizontal line with the lengths of three of the sides written above the line and the fourth below. As it happens, a quadrilateral with the specified dimensions is a geometrical im-possibility, but this cannot be said for the quadrilateral in o5, which is illustrated by a horizontal line in exactly the same manner. Whoever created such diagrams must have had only a tenuous grasp of the spatial meaning of the geometrical terms in the problems, even if he usually knew which were the right algorithms to apply to them.

10. Algorithms Used in the Problems

As we have seen, the approach to solving problems in *P.Math.* is algorithmic, that is, the student was expected to learn a sequence of arithmetical operations (additions, subtractions, multi-plications, divisions, and square roots) that was appropriate for finding the desired unknown quantity in each kind of scenario. The following survey includes both algorithms used in *P.Math.* and algorithms attested in other Greek mathematical papyri but that are related to algorithms in *P.Math.* (see also Appendix A). The repertoire ranges from the trivial (e.g., finding the area of a rectangle) to moderately complex procedures (e.g., finding an arithmetic sequence compris-ing a specified number of terms with given sum and increment). Although a respectable level of analytic reasoning underlay the more advanced algorithms, the texts give no sign that students were expected to understand why they worked.

Also characteristic of the ancient tradition of problem texts is the absence of any distinction drawn between mathematically exact and approximate algorithms. There are in fact several dif-ferent ways in which the algorithms could be inexact. All algorithms that involved the relation-ship between the diameter and circumference of a circle (P9), or the relationship between either linear measure of a circle and its area (P10A–D), involve (as we would now say) an assumed approximation for π, which was usually just 3. Hence both circumferences and areas computed from given diameters are systematically too small, whereas areas computed from given circum-ferences are systematically too large, by a little under 5 percent. In the algorithms relating to pyramids or pyramidal frusta with equilateral triangular bases (S5A and S6A), the area of an equilateral triangle is found by multiplying the square of the triangle's side by $1/3 + 1/10$, which is obviously an approximation (a good one) of $\sqrt{3}/4$. Algorithms for finding the volume or surface area of a conic frustum (i.e., a truncated right cone) required as givens the altitude of the frustum and either the diameters or the circumferences of the two circular faces, and thus had the same built-in inaccuracy arising from equating π with 3; but several of the frustum algorithms (S3A, S3B, and S4) overlay this with another approximative procedure, finding the average of the given diameters or areas and then calculating as if the object was a cylinder having that dimension, on analogy with a mathematically exact algorithm (P3A) for trapezoids. The error resulting from this short-cut would typically not be very large. For example if we are given circumferences of 1 and 2 units for the two circular faces, algorithm S3B, which employs the averaging short-cut yields a volume of 27/144 × c (where c is the altitude) whereas S3C and S3D, which assume the correct relation between the diameters and the volume, yield 28/144 × c.

The algorithm commonly employed to find the area of a quadrilateral from the given lengths of its four sides (P4, often called the "Surveyor's Formula"), also involves averaging, in this case by assuming that the area is the same as that of a rectangle whose two dimensions are the averages of each pair of opposite sides of the quadrilateral (say the average of the east and west sides times the average of the north and south sides). So long as the quadrilateral is reasonably close to being a rectangle, the algorithm yields a meaningful approximation of the true area, though for anything but a rectangle the result is always an overestimate.[22] The error becomes extreme when the quadrilateral has very acute or obtuse internal angles. For an example of how disregard of this limitation can make nonsense of a problem, see o2, O recto 8–13.

The results of arithmetical operations are generally expressed as exact quantities, not approximations. Problems involving use of the Diagonal Rule (the so-called Theorem of Pythagoras)[23] raise the potential of leading to square roots that could not be expressed as exact numbers, but in practice the people who devised the problems in the mathematical papyri showed a strong preference for "Pythagorean triangles," that is, right-angled triangles having all three sides in whole-number ratios. The 3–4–5 and 5–12–13 triangles were most commonly used, but in f6 and in *P.Bagnall* 35.1–6 we find 8–15–17 triangles. Only P.Berl. 11529v 1–10 and 11–19 have right-angled triangles with irrational hypotenuses, respectively 8–10–2√41 (approximated fairly accurately as 12 2/3 1/15 1/26 1/30) and 7–9–√130 (approximated rather crudely as 11 1/2).

Plane figures.

P1. Diagonal rule.
Given a rectangle with sides *a* and *b* (Fig. 6, top) or a right-angled triangle with legs *a* and *b* (Fig. 6, bottom), to find the diagonal or hypotenuse *c*:

    (i) Multiply *a* by itself.
    (ii) Multiply *b* by itself.
    (iii) Add the product obtained in (i) to the product obtained in (ii).
    (iv) *c* is the square root of the sum.

---

22. For a proof see Pottage 1974, 302 note 4; and for further discussion of the accuracy of the formula, Tou 2014.
23. We follow recent historiography of mathematics in antiquity (e.g., Friberg 2007, 200 or Britton, Proust, and Shnider 2011, 521) in preferring the name "diagonal rule" to "theorem of Pythagoras," partly to avoid anachronism but also to emphasize that, in the context of non-deductive mathematical problems, the relation is primarily an algorithmic means of obtaining numerical results.

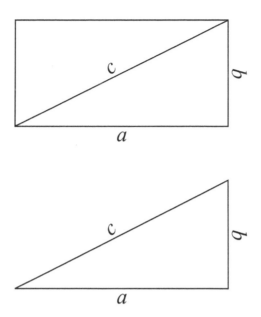

Fig. 6. Diagram for algorithms P1 and P1i.

$c = \sqrt{(a^2 + b^2)}$.

Not used directly in *P.Math.*, but cf. f6.
Other papyri: P.Berl. 11529.1–10 and 11–19; *P.Gen.* 3.124.26–40.
Demotic papyri: P.Cairo J.E. 89127-30+89137-43.N12–22 and O19-P9 (problems 30–31 and 34–35 in Parker 1972).

P1i. Inverse diagonal rule.
Given a rectangle with one side *a* and diagonal *c* (Fig. 6, top) or a right-angled triangle with one leg *a* and hypotenuse *c* (Fig. 6, bottom), to find the remaining side or leg *b*:
    (i) Multiply *c* by itself.
    (ii) Multiply *a* by itself.
    (iii) Subtract the product obtained in (ii) from the product obtained in (i).
    (iv) *b* is the square root of the difference.

$b = \sqrt{(c^2 - a^2)}$.

Used in: a5, cf. f6.
Other papyri: *P.Chic.* 3.3.16–20 (problem 5); *P.Bagnall* 35.2.23–37; *P.Gen.*3.124.1–10 and 11–25.
Demotic papyri: P.Cairo J.E. 89127-30+89137-43.M1-N11 (problems 24–29 in Parker 1972).

P2. Rectangle area algorithm.
Given a rectangle with sides *a* and *b*, to find the area *A*:

(i) A is a multiplied by b.

$$A = a \times b.$$

Used in: b2, f7.
Other papyri: P.Berl. 11529v.1.1–10; Chester Beatty Codex AC. 1390.2.14–27; *MPER* NS 1.1.13.1–4 (problem 30).

P2i. Inverse rectangle area algorithm.
Given a rectangle with side a and area A, to find the other side b:
  (i) b is A divided by a.

$$b = A \div a.$$

Not used in *P.Math*.
Other papyri: *P.Mich.* 3.151.3.6.5–8; *SB* 16.12680 verso 7–10 and 11–16.

P3A. Trapezoid area algorithm A.
Given a trapezoid having parallel sides a and b and altitude c (Fig. 7), to find the area A:
  (i) Add a and b.
  (ii) Divide sum by 2.
  (iii) A is the quotient multiplied by c.

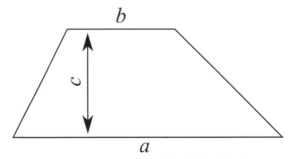

Fig. 7. Diagram for algorithm P3A.

$$A = ((a + b)/2) \times c.$$

Used in: a2, a3, b5.

P3B. Trapezoid area algorithm B.
Given a trapezoid having parallel sides a and b (with $a > b$) and equal oblique sides d (Fig. 8), one considers the trapezoid to be subdivided by perpendiculars dropped from the endpoints of the shorter parallel side into two right triangles flanking a rectangle. To find the area A:

(i) Subtract b from a.

(ii) Divide the difference by 2. The quotient is the base e of the two right triangles.

(iii) Subtract the result multiplied by itself from d multiplied by itself. The difference is the altitude c.

(iv) Multiply b by c. The product is the area of the rectangle.

(v) Divide the product by two. The quotient is the area of one of the triangles.

(vi) Multiply e by b.

(vii) A is the sum of the areas of the two triangles and the rectangle.

The algorithm embeds the inverse diagonal rule (P1i).

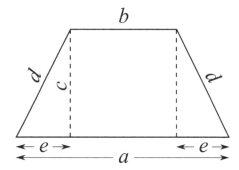

Fig. 8. Diagram for algorithm P3B.

$c = \sqrt{(d^2 - ((a - b)/2)^2)}$

$A = (((a - b)/2) \times c) + (b \times c)$.

Used in: a5.

Other papyri: presumably *P.Chic.* 3.2.1–2 (problem 3).

P3C. Trapezoid area algorithm C.

Given a trapezoid having parallel sides a and b (with $a > b$) and oblique sides d and e (with $d > e$, Fig. 9), one considers the trapezoid to be subdivided by perpendiculars dropped from the endpoints of the shorter parallel side into two right triangles flanking a rectangle. To find the area A:

(i) Multiply e by itself.

(ii) Multiply d by itself.

(iii) Subtract the square of e from the square of d.

(iv) Subtract b from a.

(v) Divide the difference into the difference of squares found in (iii).

(vi) Subtract the quotient from the difference found in (iv).

(vii) Divide the difference by 2. This is the base of the smaller right-angled triangle.

(viii) Multiply the quotient by itself.

(ix) Subtract the product from the square of e.

(x) The altitude c is the square root of the difference.

(xi) Multiply c by the base of the smaller right-angled triangle found in (vii).

(xii) Divide the difference by 2. This is the area of the smaller right-angled triangle, A_T.

(xiii) Multiply c by b. This is the area of the rectangle, A_R.

(xiv) Multiply c by the quotient found in (v) and divide the product by 2. This is the area of the scalene triangle that is the difference between the larger and the smaller right-angled triangle, A_S.

(xv) A is the sum of twice A_T, A_R, and A_S.

The algorithm embeds the inverse diagonal rule (P1i).

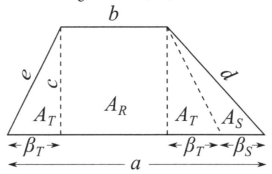

Fig. 9. Diagram for algorithm P3C.

$$\beta_S = (d^2 - e^2)/(a - b)$$
$$\beta_T = [(a - b) - \beta_S]/2$$
$$c = \sqrt{(e^2 - \beta_T^2)}$$
$$A_T = (c \times \beta_T)/2$$
$$A_R = (c \times b)$$
$$A_S = (c \times \beta_S)/2$$
$$A = 2A_T + A_R + A_S$$

Not used in *P.Math.*

Other papyri: *P.Chic.* 3.2.3–15 (problem 4); *P.Bagnall* 35.1.1–16 (variant).

P3D. Trapezoid area algorithm D.

Given an obtuse trapezoid having parallel sides a and b and oblique sides d and e (with $d > e$, Fig. 10), one considers the trapezoid to be subdivided by perpendiculars dropped from the endpoints of the shorter parallel side into two right triangles flanking a rectangle. To find the area A:

(i) Multiply e by itself.

(ii) Multiply d by itself.

(iii) Subtract the square of e from the square of d.

(iv) Subtract b from a.

(v) Divide the difference into the difference of squares found in (iii).
(vi) Subtract the difference found in (iv) from the quotient.
(vii) Divide the difference by 2. This is the base of the smaller right-angled triangle.
(viii) Multiply the quotient by itself.
(ix) Subtract the product from the square of e.
(x) The altitude c is the square root of the difference.
(xi) Subtract the base of the smaller right-angled triangle obtained in (vii) from b. This is the base of the rectangle.
(xii) Subtract the difference from a. This is the base of the larger right-angled triangle.
(xiii) Multiply c by the base of the smaller right-angled triangle obtained in (vii).
(xiv) Divide the product by 2. This is the area of the smaller right-angled triangle, A_{T1}.
(xv) Multiply c by the base of the rectangle obtained in (xi). This is the area of the rectangle, A_R.
(xvi) Multiply c by the base of the larger right-angled triangle obtained in (xii) and divide the product by 2. This is the area of the larger right-angled triangle, A_{T2}.
(xvii) A is the sum of A_{T1}, A_R, and A_{T2}.

The algorithm embeds the inverse diagonal rule (P1i).

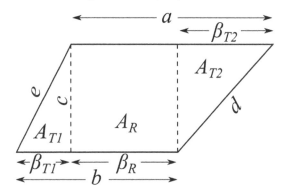

Fig. 10. Diagram for algorithm P3D.

$$\beta_{T1} = [(d^2 - e^2)/(a - b) - (a - b)]/2$$
$$c = \sqrt{(e^2 - \beta_{T1}{}^2)}$$
$$\beta_R = b - \beta_{T1}$$
$$\beta_{T2} = a - \beta_R$$
$$A_{T1} = (c \times \beta_{T1})/2$$
$$A_R = (c \times \beta_R)$$
$$A_{T2} = (c \times \beta_{T2})/2$$
$$A = A_{T1} + A_R + A_{T2}$$

Not used in *P.Math.*
Other papyri: *P.Chic.* 3.3.1–15; *P.Bagnall* 35.2.6–21 (variant).

P4. Quadrilateral area algorithm (approximate).
Given a quadrilateral having pairs of sides *a* opposite *b* and *c* opposite *d* (Fig. 11), to find the area *A*:

        (i) Add *a* and *b*.
        (ii) Divide the sum by 2.
        (iii) Add *c* and *d*.
        (iv) Divide the sum by 2.
        (v) Multiply the quotient by the quotient obtained in (ii).

The algorithm is exact only in the trivial case when the quadrilateral is a rectangle.

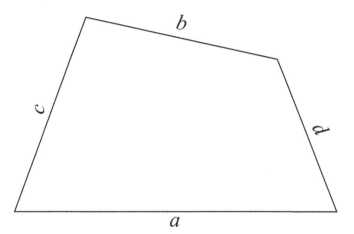

Fig. 11. Diagram for algorithms P4, P4iA, and P4iB.

$$A = ((a + b)/2) \times ((c + d)/2).$$

Used in: c1, f5, o5.
Other papyri: P.Berl. 11529v.1.1–10 (for a rectangle!); P.Col. inv. 157a.A.1–11 and B.1–16 (for a rectangle!); Chester Beatty Codex AC. 1390.1.2–14 and 2.14–27 (for a rectangle!).
Demotic papyri: BM 10520.G1–13 (for a rectangle!, problems 64–65 in Parker 1972).

P4iA. Inverse quadrilateral area algorithm A (approximate).
Given a quadrilateral with side *a* opposite side *b*, side *c*, and area *A* (Fig. 11), to find the remaining side *d*:

        (i) Multiply *A* by 2.
        (ii) Divide the product by half the sum of *a* and *b*.
        (iii) Subtract *c* from the quotient to obtain *d*.

$$d = 2A/((a + b)/2) - c.$$

Used in: o2.

P4iB. Inverse quadrilateral area algorithm B (approximate).
Given a quadrilateral with area A and side a (Fig. 11), to find the remaining sides b (opposite a), c, and d:

> (i) Factorize A into $e \times f$.
> (ii) Multiply e by 2.
> (iii) Subtract a; the difference is b.
> (iv) Let c and d both equal f.

Not used in *P.Math*.
Other papyri: Chester Beatty Codex AC. 1390.3.1–12.

P5. Hollow rectangle area algorithm.
Given a hollow rectangle having outer perimeter a, inner perimeter b, and constant perpendicular distance between outer and inner perimeter c (Fig. 12), to find the area A:

> (i) Add the outer and inner perimeters.
> (ii) Divide the sum by 2.
> (iii) A is the product of the quotient and c.

This is algorithmically homologous to trapezoid area algorithm A (P3A), and may be derived from that algorithm by subdividing the hollow rectangle into four trapezia by the straight lines joining corresponding corners of the two perimeters.

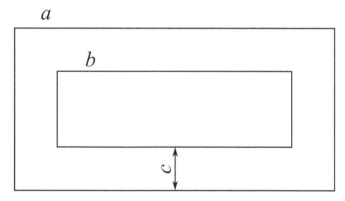

Fig. 12. Diagram for algorithm P5.

$$A = ((a + b)/2) \times c.$$

Used in: b3.

P6. Triangle area algorithm.
Given a triangle with base a and altitude c (Fig. 13), to find the area A:

> (i) Multiply a by c.
> (ii) A is the product divided by 2.

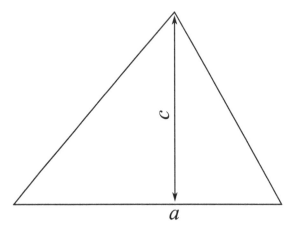

Fig. 13. Diagram for algorithm P6.

$A = (a \times c)/2.$

Used in: c3, d1, g3, n3?
Other papyri: P.Berl. 11529v.2.21–32; *MPER* NS 1.1.7.12–16 (problem 16).

P7. Right-angled triangle area algorithm.
Given a right-angled triangle with legs *a* and *b* (Fig. 14), to find the area *A*:

   (i) Multiply *a* by *b*.

   (ii) *A* is the product divided by 2.

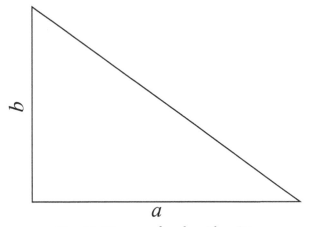

Fig. 14. Diagram for algorithm P7.

$A = (a \times b)/2.$

Used in: d3.
Other papyri: P.Berl. 11529v.1.11–19; *P.Chic.* 3.3.16–20 (problem 5).

P8A. Isosceles triangle sides algorithm A.

Given an isosceles triangle with base *a* and altitude *c* (Fig. 15), to find the equal sides *d*:

    (i) Multiply *c* by itself.

    (ii) Divide *a* by 2.

    (iii) Multiply the quotient by itself.

    (iv) Add the product obtained in (i) to the product obtained in (iii).

    (v) *d* is the square root of the sum.

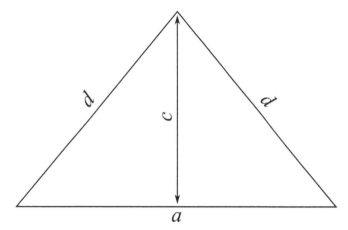

Fig. 15. Diagram for algorithms P8A and P8B.

$$d = \surd(c^2 + (a/2)^2).$$

Used in: c3.

P8B. Isosceles triangle vertical algorithm B.

Given an isosceles triangle with base *a* and equal sides *d* (Fig. 15), to find the altitude *c*:

    (i) Divide *a* by 2.

    (ii) Multiply the quotient by itself.

    (iii) Multiply *d* by itself.

    (iv) Subtract the product obtained in (ii) from the product obtained in (iii).

    (v) *c* is the square root of the difference.

The algorithm embeds the inverse diagonal rule (P1i).

$$c = \surd(d^2 - (a/2)^2).$$

Used in: d1, g3.

Other papyri: P.Berl. 11529v.2.21–32.

P9. Circle circumference algorithm (approximate).
Given a circle having diameter d (Fig. 16), to find the circumference c:
    (i) Multiply d by 3. The product is the circumference.

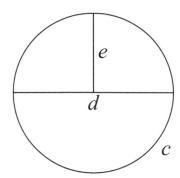

Fig. 16. Diagram for algorithms P9, P9i, P10A, P10Ai, P10B, P10Bi, P10C, and P10D.

$c = 3 \times d$.

The algorithm effectively assumes that $\pi = 3$.

Not used in *P.Math.*
Other papyri: BM Add. MS 41203A.r.1–6; *P.Oxy.* 3.470.31–46.

P9i. Circle inverse circumference algorithm (approximate).
Given a circle having circumference c (Fig. 16), to find the diameter d:
    (i) Divide c by 3. The quotient is the diameter.

$d = c \div 3$.

The algorithm effectively assumes that $\pi = 3$.

Not used in *P.Math.*
Other papyri: *MPER* NS 15.172–174.1–6.

P10A. Circle area algorithm A (approximate).
Given a circle having diameter d (Fig. 16), to find the area A:
    (i) Multiply d by itself.
    (ii) Subtract one quarter of the product from itself. The difference is A.

$A = (d^2 - d^2/4) = 3(d^2)/4$.

Since the exact formula is $A = \pi(d^2)/4$, the algorithm effectively assumes that $\pi = 3$.

Used in: e1.
Other papyri: *MPER* NS 1.1.9a.1–2 (problem 20), 9a.2–5 (problem 21), 9a.5–6 (problem 22), 9a.6–7 (problem 23); *MPER* NS 15.178.3.1–6; *PSI* 3.186v.1–8.

P10Ai. Inverse circle area algorithm A (approximate).
Given a circle having area A (Fig. 16), to find the diameter d:
      (i) Add to A one-third of itself.
      (ii) d is the square root of the result.

$$d = \sqrt{(A + A/3)} = \sqrt{(4A/3)}.$$

Used in: d4.
Demotic papyri: P.Cairo J.E. 89127-30+89137-43.O1-18 (problems 32–33 in Parker 1972).

P10B. Circle area algorithm B (approximate).
Given a circle having circumference c (Fig. 16), to find the area A:
      (i) Multiply c by itself.
      (ii) Divide the result by 12. The quotient is A.

$$A = (c^2)/12.$$

The exact formula is $A = (c^2)/4\pi$, so again the algorithm assumes $\pi = 3$.

Used in: g2 (incorrectly).
Other papyri: *MPER* NS 15.178.2.1–7; *MPER* NS 15.172–174.1–6 (14 instead of 12 in step ii is surely a textual error).

P10Bi. Inverse circle area algorithm B (approximate).
Given a circle having area A (Fig. 16), to find the circumference c:
      (i) Multiply A by 12.
      (ii) c is the square root of the product.

$$c = \sqrt{(12A)}.$$

Used in: e2.

P10C. Circle area algorithm C (approximate).
Given a circle having diameter d and perpendicular half-diameter e (Fig. 16), to find the area A:

(i) Add d and e.
(ii) Multiply the sum by itself.
(ii) Divide by 3. The quotient is A.

$A = (d + e)^2/3$.

The algorithm effectively assumes that $\pi = 3$.

Not used in *P.Math.*
Other papyri: *MPER* NS 15.178.4.1–9.

P10D. Semicircle area algorithm D (approximate).

Given a semicircle having diameter d and perpendicular half-diameter e (Fig. 16), to find the area A:
    (i) Add d and e.
    (ii) Divide by 2.
    (ii) Multiply by e. The quotient is A.

$A = e(d + e) / 2$.

The algorithm effectively assumes that $\pi = 3$.

Not used in *P.Math.*
Other papyri: *MPER* NS 15.178.5.1–9.

Solids.

S1. Parallelepiped volume algorithm.

Given a rectangular parallelepiped with linear dimensions a, b, c (Fig. 17), to find the volume V:
    (i) Multiply a by b.
    (ii) V is the product multiplied by c.

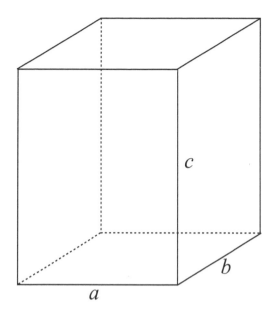

Fig. 17. Diagram for algorithm S1.

$$V = a \times b \times c.$$

Used in: c5, g4, i2.
Other papyri: Chester Beatty Codex AC. 1390.2.1–6; *P.Cair. cat.* 10758.3v1.8–12 (problem 2); *P.Lond.* 5.1718.4.71; *T.Varie* 20.3–4 and 5–6; *MPER* 1.1.2.4–7 (problem 2), 2.8–13 (problem 3), 3.7–12 (problem 5), 3.12–15 (problem 6), and 4.3–8 (problem 8).

S2. Prism volume algorithm.
Given a prism (or cylinder) having cross section area A and perpendicular dimension h (Fig. 18), to find the volume V:

(i) V is A multiplied by h.

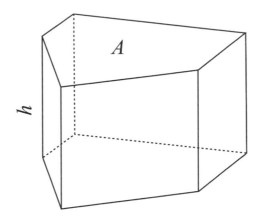

Fig. 18. Diagram for algorithms S2 and S2i.

$$V = A \times h.$$

Used in: a2, a3, b3, b5, c1, e1, g2, n3, n2.
Other papyri: Chester Beatty Codex AC. 1390.1.2–14; *PSI* 3.186v.1–8 (cylinder); *MPER* NS 1.1.7.12–16 (problem 16), 9a1–2 (problem 20), 9a2–5 (problem 21), 9a5–6 (problem 22), and 9a6–7 (problem 23).

S2i. Inverse prism volume algorithm.
Given a prism having volume V and perpendicular dimension h (Fig. 18), to find the area of the cross section A:

    (i) A is V divided by h.

$$A = V/h.$$

Used in: d4, e2.

S3A. Conic frustum volume algorithm A (approximate).
Given a conic frustum having circular faces with diameters a_1 and a_2 and perpendicular dimension c (Fig. 19), to find the volume V:

    (i) Add a_1 and a_2.
    (ii) Divide the sum by 2.
    (iii) Multiply the sum by itself.
    (iv) Subtract a quarter of the result from itself.
    (v) V is the difference multiplied by c.

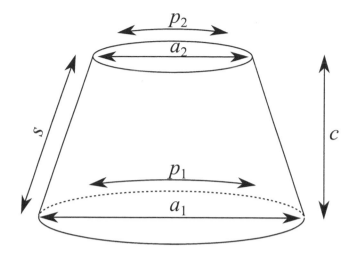

Fig. 19. Diagram for algorithms S3A, S3B, S3C, S3D, and S4.

$$V = (d^2 - d^2/4) \times c, \text{ where } d = (a_1 + a_2)/2.$$

The algorithm embeds Circle Algorithm A (P10A) with its implied value of 3 for π, and moreover calculating the frustum's volume as if of a cylinder with diameter equal to the average of a_1 and a_2 is only an approximation.

Used in: n1.
Other papyri: BM Add. MS 41203.11–16 (erroneously treating circumferences as diameters).
Demotic papyri: BM 10399.B1–C29 (problems 42–45 in Parker 1972).

S3B. Conic frustum volume algorithm B (approximate).
Given a conic frustum having circular faces with circumferences p_1 and p_2 and perpendicular dimension c (Fig. 19), to find the volume V:
     (i) Add p_1 and p_2.
     (ii) Divide the sum by 2.
     (iii) Multiply the sum by itself.
     (iv) Divide the product by 12.
     (v) V is the difference multiplied by c.

$$V = d^2 \times c /12, \text{ where } d = (p_1 + p_2)/2.$$

Not used in *P.Math.*
Other papyri: Chester Beatty Codex AC. 1390.1.15–23; *P.Cair. cat.* 10758.3v1.1–7 (problem 1, but for the division by 36 instead of 12 in step iv cf. S3C).

S3C. Conic frustum volume algorithm C (approximate).

The application of this algorithm in Greek mathematical papyri is a conjecture based on the divisions by 36 cited below.[24]

Given a conic frustum having circular faces with circumferences p_1 and p_2 and perpendicular dimension c (Fig. 19), to find the volume V:

(i) Multiply p_1 by itself.
(ii) Multiply p_2 by itself.
(iii) Multiply p_1 by p_2.
(iv) Add the three products.
(v) Multiply the sum by c.
(v) V is the product divided by 36.

$$V = (p_1{}^2 + p_1 \times p_2 + p_2{}^2) \times c/36$$

Not used in *P.Math.*

Other papyri: Possibly *P.Lond.* 5.1718.4.75–76, and cf. the division by 36 instead of 12 in *P.Cair. cat.* 10758.3v1.1–7 (problem 1).

S3D. Conic frustum volume algorithm D (approximate).

Given a conic frustum having circular base and top with circumferences p_1 and p_2, and slope height (measured along the conical surface) s (Fig. 19), to find the volume V:

(i) Divide p_1 by 3. The quotient is the diameter of the base, a_1.
(ii) Divide p_2 by 3. The quotient is the diameter of the top, a_2.
(iii) Subtract a_2 from a_1.
(iv) Divide the difference by 2.
(v) Multiply the quotient by itself.
(vi) Multiply s by itself.
(vii) Subtract the product obtained in step (v) from the product obtained in step (vi).
(viii) Find the square root of the difference. This is the altitude c.
(ix) Add a_1 and a_2.
(x) Divide the sum by 2.
(xi) Multiply the quotient by itself.
(xii) Subtract a quarter of the result from itself.
(xiii) Subtract a_2 from a_1.
(xiv) Divide the difference by 2.
(xv) Multiply the quotient by itself.
(xvi) Divide the product by 4.
(xvii) Add the results obtained in steps (xii) and (xvi).
(xviii) V is the sum multiplied by c.

24. Smyly 1920, 107; Kurt Vogel in MPER NS 1, 39 doubts that an exact algorithm taking this form existed in antiquity.

$a_1 = p_1/3$
$a_2 = p_2/3$
$c = \sqrt{(s^2 - d_1{}^2)}$ where $d_1 = (a_1 - a_2)/2$
$V = ((d_2{}^2 - d_2{}^2/4) + d_1{}^2/4) \times c$, where $d_2 = ((a_1 + a_2)/2)$

Not used in *P.Math.*
Other papyri: *MPER* NS 1.1.9b.1–9 (problem 24) and 10.7–16 (problem 25).

S4. Conic frustum surface algorithm (approximate).

Given a conic frustum having circular faces with diameters a_1 and a_2 and perpendicular dimension c, to find the surface A:

  (i) Add a_1 and a_2.
  (ii) Divide the sum by 2.
  (iii) Multiply the sum by 3.
  (iv) V is the product multiplied by c.

$$V = 3 \times (a_1 + a_2)/2 \times c$$

Not used in *P.Math.*
Other papyri: BM Add. MS 41203A.r.1–6; *P.Oxy.* 3.470.31–46.

S5A. Pyramid volume algorithm A.

Given a pyramid having an equilateral triangle with side a as base and similar equilateral triangles with sloping side s as the other faces (Fig. 20), to find the volume V:

  (i) Multiply a by itself.
  (ii) Divide the product by 3. The quotient is the *epipedos* (square of a line from the base's center to a vertex of the base).
  (iii) Multiply s by itself.
  (iv) Subtract the *epipedos* from the product.
  (v) Find the square root of the difference. The result is the vertical, c.
  (vi) Multiply the square of a (from step i) by 1/3 1/10. The product is the area of the base, A.
  (vii) Multiply A by c.
  (viii) V is the product divided by 3.

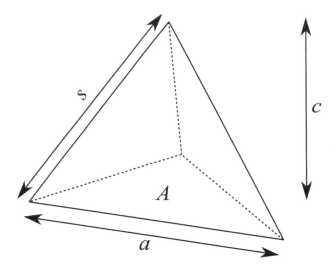

Fig. 20. Diagram for algorithm S5A.

$c = \sqrt{(s^2 - a^2/3)}$
$A = 13/30 \times a^2$
$V = A \times c/3$

Not used in *P.Math*.
Other papyri: *MPER* NS 1.1.5.1–5 (problem 10), 5.19–6.3 (problem 12), and 11.10–18 (problem 27).

S5B. Pyramid volume algorithm B.
Given a pyramid having a square with side *a* as base and similar equilateral triangles with sloping side *s* as the other faces (Fig. 21), to find the volume *V*:

    (i) Divide *a* by 2.
    (ii) Multiply the quotient by itself.
    (iii) Add the product to itself. The sum is the *epipedos* (square of a line from the base's center to a vertex of the base).
    (iv) Multiply *s* by itself.
    (v) Subtract the *epipedos* from the product.
    (vi) Find the square root of the difference. The result is the vertical, *c*.
    (vi) Multiply *a* by itself. The product is the area of the base, *A*.
    (vii) Multiply *A* by *c*.
    (viii) *V* is the product divided by 3.

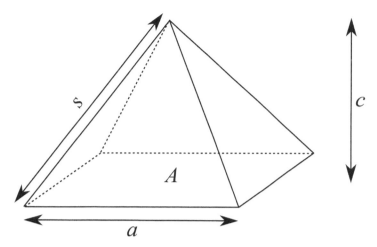

Fig. 21. Diagram for algorithm S5B.

$c = \sqrt{(s^2 - a^2/2)}$
$A = a^2$
$V = A \times c/3$

Not used in *P.Math.*
Other papyri: *MPER* NS 1.1.8.1–12 (problem 18) and 12.8–13 (problem 29).

S5B. Pyramid volume algorithm C.
Given a pyramid having base with area A and vertical c, to find the volume V:
> (i) Multiply A by c.
> (ii) V is the product divided by 3.

$V = A \times c/3$

Not used in *P.Math.*
Other papyri: *MPER* NS 1.1.7.9–12 (problem 15) and 13.1–4 (problem 30).
Demotic papyri: P.Cairo J.E. 89127-30+89137-43.S18-23 (problem 40 in Parker 1972).

S6A. Pyramidal frustum volume algorithm A.
Given a pyramidal frustum having equilateral triangles with sides a_1 and a_2 as base and top, and similar equilateral triangles with sloping side s as the other faces (Fig. 22), to find the volume V:
> (i) Subtract a_2 from a_1.
> (ii) Multiply the difference by itself.
> (iii) Divide the product by 3.
> (iv) Multiply s by itself.

(v) Subtract the quotient obtained in step (iii) from the product.

(vi) Find the square root of the difference. The result is the vertical, c.

(vii) Add a_1 and a_2.

(viii) Divide the sum by 2.

(ix) Multiply the quotient by itself.

(x) Multiply the product by 1/3 1/10.

(xi) Divide the difference of a_1 and a_2 (from step i) by 2.

(xii) Multiply the quotient by itself.

(xiii) Multiply the product by 1/3 1/10.

(xiv) Divide the product by 3.

(xv) Add the results obtained in steps (x) and (xiv).

(xvi) V is the sum multiplied by c.

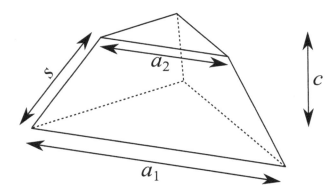

Fig. 22. Diagram for algorithm S6A.

$$c = \sqrt{(s^2 - (a_1 - a_2)^2/3)}$$
$$V = 13/30 \times (((a_1 + a_2)/2)^2 + ((a_1 - a_2)/2)^2 / 3) \times c$$

Not used in *P.Math*.

Other papyri: *MPER* NS 1.1.5.6–18 (problem 11), 6.3–11 (problem 13), 7.1–8 (problem 14), and 11.1–10 (problem 26).

S6B. Pyramidal frustum volume algorithm B.

Given a pyramidal frustum having squares with sides a_1 and a_2 as base and top, and similar equilateral triangles with sloping side s as the other faces (Fig. 23), to find the volume V:

(i) Subtract a_2 from a_1.

(ii) Multiply the difference by itself.

(iii) Divide the product by 2.

(iv) Multiply s by itself.

(v) Subtract the quotient obtained in step (iii) from the product.

(vi) Find the square root of the difference. The result is the vertical, c.

(vii) Add a_1 and a_2.

(viii) Divide the sum by 2.
(ix) Multiply the quotient by itself.
(x) Divide the difference of a_1 and a_2 (from step i) by 2.
(xii) Multiply the quotient by itself.
(xiii) Divide the product by 3.
(xiv) Add the results obtained in steps (ix) and (xiii).
(xv) V is the sum multiplied by c.

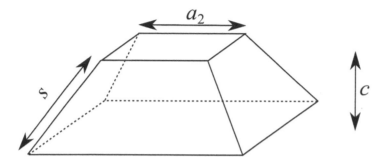

Fig. 23. Diagram for algorithm S6B.

$c = \sqrt{(s^2 - (a_1 - a_2)^2/2)}$
$V = 13/30 \times (((a_1 + a_2)/2)^2 + ((a_1 - a_2)/2)^2/3) \times c$

Not used in *P.Math.*
Other papyri: *MPER* NS 1.1.8.12–19 (problem 19) and 12.1–8 (problem 28).

Numbers.

N1. Arithmetical sequence algorithm.
It is required to find n numbers $i_1 \dots i_n$ increasing by constant differences d, and such that their sum is s. To find the smallest of the numbers i_1:

(i) Multiply n by itself.
(ii) Add n to the product.
(iii) Divide the sum by 2.
(iv) Multiply the quotient by d.
(v) Subtract the product from s.
(vi) Divide the difference by n.
(vi) Add d to the quotient. The sum is i_1.

Successive additions of d yield the other numbers.

$$i_1 = (s - ((n^2 + n)/2) \times d)/n + d.$$

Used in: m2.

11. Partitions into Unit-Fractions

Six of *P.Math.*'s mathematical problems (a4, c4, f3, f4, and the very poorly preserved h2 and i4) are of a special type, with elliptical statement of the problem λοιπαὶ *R*, τὸ *D* ἐν *n* μόρια· μὴ πρόβα ρ̄ (with *R*, *D*, and *n* standing for numerals), which we believe should be translated, "remainder *R*, the *D*th part in *n* unit fractions; do not surpass 100." This is immediately followed by the answer, taking the form ἔσται τὰ μόρια followed by a sequence of numerals to be read as unit-fractions, always in order of increasing denominators. The meaning of such a problem is as follows. It is to be supposed that a division of some number by *D* has been carried out, such that the remainder beyond a whole number of *D*s is *R*. We are now required to divide *R* by *D* and express the resulting fractional quotient $R \div D$ as a sum of *n* distinct unit-fractions, none of whose denominators should exceed 100. (What the original dividend or the integer part of the quotient were is immaterial to the problem.) In contrast to the other problems, no steps leading to the answer are provided.

P.Yale 4.187 (forthcoming) consists of comparable fraction partitions in a didactic context, for example i 6: ια, τὸ ξθ, ιβ κγ μϛ ϙβ, to be interpreted as "11, the 69th part (is) 1/12 1/23 1/46 1/92." The thirty-four such partitions do not constitute a reference table; notwithstanding a few groups of consecutive lines having the same dividend, including a sequence (ii 15–iii 22) dividing the same quantity by 10 through 17, there is no broad pattern in the numbers. The partitions can be into as few as two or as many as five unit-fractions, none of which has a denominator greater than 98, so it seems that there was an unstated restriction to denominators not exceeding 100.

Another manuscript that contains comparable problems is *P.Cair. cat.* 10758. In *P.Cair. cat.* 10758, problems 18, 21, 22, 23, 38, 39, and 40 are stated with an algorithm such as τῆς *R* τὸ *D*, "of *R* the *D*th part," and the solution (incomplete in 40) is typically given as ὡς εἶναι, "so that they are," followed by a sequence of unit-fractions, with denominators never exceeding 99, that sum to $R \div D$. Moreover, *P.Cair. cat.* 10758 problems 16, 19, 20, and 50 are similar but also include a stipulation of the number of unit-fractions, e.g., χώρισον *D* εἰς *n* μόρια, "partition the *D*th part into *n* unit-fractions," such as we have in the *P.Math.* problems. In contrast to *P.Math.*, *P.Cair. cat.* 10758 provides a step-by-step solution of each problem. These solutions are made up of several recurring types of operations:

- Conversion of a division to lowest whole-number terms $R \div D$ (i.e., *R* and *D* are whole numbers with no common factor).
- Scaling up of a division $R \div D$ by a whole-number multiple *k* to obtain $(kR) \div (kD)$.
- Peeling off of a unit-fraction from a division in whole number terms R/D, such that $E \times F = D$, obtaining the unit-fraction $1/F$ plus the remainder $(R-E) \div D$.
- Peeling off of a unit-fraction from a division in whole-number terms $R \div D$, obtaining the unit-fraction $1/D$ plus the remainder $(R-1) \div D$. [This is a special case of the preceding operation.]

- Splitting of a division in whole-number terms $R÷D$ into two unit-fractions $1/(E×[E+F]÷R)$ and $1/(F×[E+F]÷R)$ such that $E×F = D$ and $E+F$ is divisible by R.
- Splitting a single unit-fraction $1/D$ into two unit-fractions $1/(E×[E+F])$ and $1/(F×[E+F])$ such that $E×F = D$. [This is a special case of the preceding operation.]

For example, *P.Cair. cat.* 10758 problem 19 requires us to express a quantity given as the sum of the three unit fractions $1/55 + 1/56 + 1/70$ as a sum of four unit-fractions. The solution is as follows:

i. Convert $1/55 + 1/56 + 1/70$ to a division in lowest whole number terms: $31÷616$.
ii. Peel off the unit-fraction $1/88$ (i.e. $7÷616$), leaving remainder $24÷616$.
iii. Convert to lowest terms: $3÷77$.
iv. Peel off the unit-fraction $1/77$, leaving remainder $2÷77$.
v. Split into two unit-fractions $1/63$ and $1/99$ (using $77 = 7 \times 11$).
vi. Answer: $1/63 + 1/77 + 1/88 + 1/99$.

As this example illustrates, problems of this kind are puzzles with no immediate practical application. Why would one want to express a given quantity as a sum of a predetermined number of unit-fractions, especially when we know from the start that it can be expressed in fewer? In the foregoing problem we are given the quantity as a sum of three unit-fractions and asked to re-express it as a sum of four; in others, the given quantity already is, or can be simplified to, a single unit-fraction.[25] Secondly, such problems could not be solved mechanically by applying a single memorized algorithm; the approach is exploratory, with arbitrary choices such as whether to scale up a division and if so by what multiple, and which of multiple factorizations to use in a splitting operation. For example a division $1÷24$ can be split as $1/25 + 1/600$ or as $1/28 + 1/168$ or as $1/33 + 1/88$ or as $1/40 + 1/60$, depending on which factorization of 24 one chooses. The prohibition of denominators greater than 100 in *P.Math.*'s problems would rule out some of these possibilities, and its intention may have been to make the problems more challenging, in particular by eliminating the lazy option of splitting any unit-fraction $1/R$ into $1/(R+1) + 1/R(R+1)$. The partitions offered in both *P.Cair. cat.* 10758 and *P.Math.* tend to consist of unit-fractions whose denominators are of roughly the same order of magnitude.

Although the problem of partitioning a quantity into say four unit-fractions might be approached in multiple ways, for example by an initial splitting into two parts, each of which is then split into a pair of unit-fractions, the solutions in *P.Cair. cat.* 10758 exhibit a linear approach, comprising a sequence of peelings-off terminating in a splitting. The denominators of

25. In special circumstances there is a point in splitting a single unit-fraction into two. Specifically, splitting an odd-numbered unit-fraction into two even-numbered ones makes it easy to express double the fraction in distinct unit-fractions.

the unit-fractions resulting from a splitting have the property that their sum is either a square or a small multiple of a square. This fact provides a clue to reconstructing the routes by which the solutions in *P.Math.* were obtained. On the other hand, unit-fractions peeled off at the beginning tend to share factors with the given divisor; for example in *P.Cair. cat.* 10758 problem 19, 616 is 7 × 88, and the first unit-fraction peeled off is 1/88.

The following conjectural reconstructions of *P.Math.*'s partition problems may not be exactly how the originators of these problems proceeded, but we suspect that their approaches must have been along similar lines. Problem f3 turns out to be fairly challenging.

Conjectural reconstructions of the partition problems in P.Math.

a4. 4 ÷ 80, partition into four unit-fractions.

i.	Convert to lowest terms: 1 ÷ 20.
ii.	Scale up by 7: 7 ÷ 140.
iii.	Peel off 2 ÷ 140, i.e. 1/70, leaving remainder 5 ÷ 140.
iv.	Convert to lowest terms: 1 ÷ 28.
v.	Scale up by 3: 3 ÷ 84.
vi.	Peel off 1/84, leaving remainder 2 ÷ 84.
vii.	Convert to lowest terms: 1 ÷ 42.
viii.	Split into two unit-fractions 1/78 and 1/91 (using 42 = 6 × 7).
ix.	Answer: 1/70 + 1/78 + 1/84 + 1/91.

c4. 1 1/4 ÷ 7, partition into seven unit-fractions.

i.	Convert to lowest terms: 5 ÷ 28.
ii.	Peel off 1/28, leaving remainder 4 ÷ 28.
iii.	Convert to lowest terms: 1 ÷ 7.
iv.	Scale up by 5: 5 ÷ 35.
v.	Peel off 1/35, leaving remainder 4 ÷ 35.
vi.	Scale up by 2: 8 ÷ 70.
vii.	Peel off 1/70, leaving remainder 7 ÷ 70.
viii.	Convert to lowest terms: 1 ÷ 10.
ix.	Scale up by 6: 6 ÷ 60.
x.	Peel off 1/60, leaving remainder 5 ÷ 60.
xi.	Convert to lowest terms: 1 ÷ 12.
xii.	Scale up by 2: 2 ÷ 24.
xiii.	Peel off 1/24, leaving remainder 1 ÷ 24.
xiv.	Scale up by 2: 2 ÷ 48.
xv.	Split into 1/42 and 1/56 (using 48 = 6 × 8).
xvi.	Answer: 1/24 + 1/28 + 1/35 + 1/42 + 1/56 + 1/60 + 1/70.

f3. 9 ÷ 119, partition into four unit-fractions.

   i.   Scale up by 2: 18 ÷ 238.

   ii.   Peel off 7 ÷ 238 (i.e. 1/34), leaving 11 ÷ 238.

   iii.   Scale up by 3: 33 ÷ 714.

   iv.   Peel off 14 ÷ 714 (i.e. 1/51), leaving remainder 19 ÷ 714.

   v.   Scale up by 2: 38 ÷ 1428.

   vi.   Split into 1/68 and 1/84 (using 1428 = 34 × 42, with 34 + 42 divisible by 38).

   vii.   Answer: 1/34 + 1/51 + 1/68 + 1/84.

f4. 36 ÷ 228, partition into five unit-fractions.

   i.   Peel off 19 ÷ 228 (i.e. 1/12), leaving remainder 17 ÷ 228.

   ii.   Peel off 4 ÷ 228 (i.e. 1/57), leaving remainder 13 ÷ 228.

   iii.   Peel off 3 ÷ 228 (i.e. 1/76), leaving remainder 10 ÷ 228.

   iv.   Convert to lowest terms: 5 ÷ 114.

   v.   Split into 1/30 and 1/95 (using 114 = 6 × 19, with 6 + 19 is divisible by 5).

   vi.   Answer: 1/12 + 1/30 + 1/57 + 1/76 + 1/95.

12. Metrological Relations and Coefficients in the Mathematical Problems

Only a small subset of the units of the metrological texts appear in the problems. Among the numerous units of length, for example, only the cubit, finger, and schoinion occur frequently, and additionally we find the xylon, pace, and reed in one problem each (c5 and f5), where quantities in these units are to be converted to cubits. Again, areas are almost always expressed in arourai, just once in bikoi (o3).

Most relations assumed in problems can be found somewhere in the metrological texts. For example, f5 converts area cubits to arourai by dividing by 9216, a relation found in e3 (E verso 23–24), while in o3 area cubits are divided by 200 to obtain bikoi, as stated again in e3 (E verso 19). On the other hand, none of the metrological texts asserts the relation that one aroura is the area of a square whose side is one schoinion, which is frequently invoked in the problems; at best one might derive it roundabout by recognizing that the number of area cubits in an aroura, 9216, is the square of 96, the number of cubits in a schoinion. The liquid measure unit xestes (sextarius), in which i2 requires one to express the capacity of a vat, does not appear in the metrological texts at all, at least so far as they are preserved.

Four problems, a3, b5, c1, and g4, present a metrological enigma. Each of them concerns a long, sticklike shape, whose length is given in cubits while the dimensions defining the cross-section are given in fingers, and the assignment is to find the volume. As an intermediate step, the area of the cross-section (in square fingers) is multiplied by the length (in cubits) to obtain the volume in quasi-units corresponding to a rectangular parallelepiped one finger by one finger by one cubit. This product is then divided by either 192 (b5) or 288 (a3, c1, g4) to obtain a result in unnamed units, with any remainder in "fingers" that are one-twenty-fourth of the unnamed unit, i.e., the same fraction that a regular linear finger is of a linear cubit, from which one might expect that the unnamed unit is supposed to be the volume cubit. The diagram of g4, G verso 17, makes this identification explicit, if the word πηχος prefixed there to the numerical result represents πῆχεις, "cubits." The number of finger-by-finger-by-cubits in a volume cubit, however, is 576, so that dividing the interim result in these problems by 192 or 288 amounts to converting into units equivalent respectively to one-third or one-half of a solid cubit. The metretes is indeed one-third of a solid cubit, but it is a unit of liquid measure, whereas the problems in question are clearly about solid objects. The nonstandard system of apparently volumetric units at the end of text h1 (H →16–19) seems to offer a volumetric spithame equal to half a solid cubit, and a volumetric lichas equal to a third of a solid cubit, but even if this strange system was not just a figment of the text's writer's imagination, its fingers are twenty-fourths specifically of the volumetric cubit, so not consistent with the volumetric fingers in the problems.

Problems b3 and f7 illustrate the use of standard coefficients, that is, constants required for the conversion of a magnitude expressed in general units into a number of concrete objects potentially constituting that magnitude, specifically the number of bricks (48) making up one solid cubit, and the number of area cubits of land (4) that would be devoted to a single vine. Like metrological relations, these coefficients were presumably memorized, and they are applied in the final steps of the relevant problems without any explanation of what they are.

13. The Nature of the Codex

What is this codex, and who was responsible for it? Despite all of the information that we have elicited and compiled, the answers to these questions are not immediately obvious. We initially thought that it might be a teacher's compendium of material, but we have found this hypothesis more and more difficult to defend as we have looked carefully at its characteristics.

(1) As we have seen, we are missing part of the codex and in all likelihood are lacking the opening pages. If there were any indications of authorship or ownership in its original state—and there are often such indications in education-related texts, particularly tablets—they are lost. What remains does not have any obvious organization. If H and I were located at the front, the model documents may have been roughly grouped, but we have no clear evidence that they were, or of what might have preceded them. There is no progression in terms of subject or difficulty, either. This is certainly not a systematic compendium of any sort.

(2) The handwriting is fairly fluent and at times stylish. If the model contracts and the mathematical texts are all by the same hand but in different styles, this suggests someone with enough education to have acquired a reasonable level of skill in writing, but by no means a completely fluent documentary script. Moreover, there is a high level of inconsistency in the writing and the layout does not always make a good impression.

(3) The level of spelling errors is high, and there are many inconsistencies; they do not all trend in the same direction. Most are explicable as the normal phonetic variants found in documentary and epistolary papyri, where people are writing words down as they heard them.

There is no sign of the command of spelling that would have been acquired in a grammarian's classroom.

(4) More damagingly, there are many errors that are not phonetic, that suggest a failure to understand the text at all. Some interchanges are unparalleled in the papyri, and these are not phonetic interchanges. Some might be failures to understand a character and thus to copy it correctly. Morphology is mediocre at best, and in many instances the case endings are wrong. Again, this is a sign of a lack of any literary education above the most elementary level.

(5) Arithmetic is unfailingly correct, but the skill in computation is often applied in the service of mistaken algorithmic procedures, to the point of nonsense in some cases, as where four dimensions seem to be envisaged. The diagrams attached to the problems are often wrong or even disconnected from the givens of the problems. The writer cannot be said to have made up for a weak literary grounding with a high level of expertise in geometry.

Taking all of these characteristics together, we believe that the most likely explanation is that the codex belonged to a student in a school devoted to training business agents and similar professionals. Much of it may have been copied by eye rather than from dictation, but there are elements of both processes at different points, and the exemplar may itself have been the result of things taken down by ear. It is perfectly conceivable that a student took down information on tablets in the classroom, then made an intended "fair copy" onto papyrus at a later time. Such a process could account for the mixture of types of errors in the codex. It would also be a plausible setting for the habit visible in many places of beginning a section with some ambition for elegant presentation but gradually becoming less and less in command of the writing as time went on and line succeeded line. Such initial ambition followed by tailing off can also be seen in many private letters.

A brief account of numerical literacy is given by Raffaella Cribiore in *Gymnastics of the Mind*.[26] She notes the presence of some basic training in arithmetic in the lower schools, but points to the acquisition of deeper competence in multiplication and fractions in what she calls "specialized scribal schools." She points in particular to the relatively skilled handwriting of these "apprentice scribes" as evidence that this teaching does not belong to the earliest stages of education. Our codex fleshes out this picture importantly, suggesting that model contracts, metrological information, geometry, and some types of fractional computation all were part of an education that was indeed separate from that of the grammarian and did not presuppose any time in a grammarian's classroom.[27] At the same time, it suggests that "scribal" is not the best term for an education that aimed to inculcate a number of skills relevant for the administration of land, rents, and taxes, skills that would have been equally relevant for the agents of liturgists

---

26. Cribiore 2001, 180–183.

27. For other apparently didactic manuscripts from late antiquity combining multiple genres such as mathematical problems, arithmetical tables, metrology, and model contracts, see BL Add. MS 41203 (of which only tablet A, containing mathematical problems and isopsephisms, has been published in Skeat 1936, but a complete edition of all four tablets by Julia Lougovaya and Rodney Ast is in preparation), BL Add. MS 33369 (forthcoming, see section 6 above), and *P. Yale* 4.184–187 (forthcoming). In Add. MS 33369, Todd Hickey observes that the handwriting points to an early stage of education.

who had to assess and collect taxes and the stewards of the estates of large landowners: quite possibly the same people at different times, in fact, as the landowners were the liturgists. The precise nature of this schooling has never been studied in detail, but our codex makes a substantial contribution to the investigation of this subject.

14. Index of Texts

a1 (Ar 1–12) model document: loan of money

a2 (Ar 14–20) trapezoidal solid (breach in dike, διάκοπος ἐπὶ χώματος)

a3 (Av 1–7) trapezoidal solid

a4 (Av 8–9) partition into unit-fractions

a5 (Av 11–12, Br 1–3) trapezoid

b2 (Br 4–8) rectangle (field)

b3 (Br 10–21, Bv 1) complex solid (tower, πύργος, bricks)

b4 (Bv 2–7) rectangle

b5 (Bv 8–16) trapezoidal solid

c1 (Cr 1–12) quadrilateral prism (beam?, ξύλον)

c2 (Cr 14–21) shipload of wheat

c3 (Cv 1–9) isosceles triangle

c4 (Cv 11–12) partition into unit-fractions

c5 (Cv 13–20) rectangular solid (excavation of river, διῶρυξ ποτάμου)

d1 (Dr 1–12) isosceles triangle

d2 (Dr 13–19) speeds of runners

d3 (Dv 1–17) right-angled triangle

d4 (Dv 19–25) conical frustum (circular pit, ὄρυγμα στρογγύλον)

e1 (Er 1–10) conical frustum (circular pit, ὄρυγμα στρογγύλον)

e2 (Er 11–20) conical frustum (circular pit, ὄρυγμα στρογγύλον)

e3 (Ev 1–24) metrology: units of length, area, volume

f1 (Fr 1–5) sale of artabas of wheat

f2 (Fr 7–18) three granaries

f3 (Fr 19–21) partition into unit-fractions

f4 (Fr 22–23) partition into unit-fractions

f5 (Fv 1–9) quadrilateral (field)

f6 (Fv 11–17) right-angled triangle

f7 (Fv 18–22) rectangle (vineyard, χωρίον ἀμπέλιον, vines)

g1 (Gr 1–20) metrology: units of length

g2 (Gr 22–28) cylinder (circular *naubion*, ναύβιον στρογγύλον)

g3 (Gv 1–9) isosceles triangle

g4 (Gv 10–17) circular or rectangular solid (circular *naubion*?, [ναύβιο]ν? στρογγύλον)

m1 (Mr 1–19) metrology: units of length, area, volume, liquid capacity m2 (Mv 1–15) distribution of artabas of wheat

n1 (Nr 1–10) conical frustum (cistern?)

n2 (Nr 12–17) triangular solid (granary, θησαυρὸς τρίγωνος)

n3 (Nv 1–7) rectangular solid (granary, θησαυρός, artabas of grain)

n4 (Nv 8–20) complex solid (vaulted granary, θησαυρὸς καμαρωτός)

o1 (Or 1–7) complex solid (vaulted granary, θησαυρὸς καμαρωτός)

o2 (Or 8–13) square (field, σφραγίς)

o3 (Or 14–20) trapezoid? (vacant lot, ψιλός)

o4 (Or 22–26) triangle

o5 (Ov 1–11) quadrilateral (field?)

o6 (Ov 12–19) pay

h1 (H → 1–19) metrology: units of length

h2 (H → 21–22) partition into unit-fractions

h3 (H ↓ 1–15) model document: undertaking to lease arable land

i1 (I → 1–14) metrology: units of length, liquid capacity, weight

i2 (I → 16–23) rectangular solid (vat, ληνός)

i3 (I ↓ 1–12) model document: loan of money

i4 (I ↓ 13–14) partition into unit-fractions

i5 (I ↓ 16–18) unidentified

15. Index of Texts by Type

Model Documents.

a1 (Ar 1–12) loan of money

h3 (H ↓ 1–15) undertaking to lease arable land

i3 (I ↓ 1–12) loan of money

Metrology.

e3 (Ev 1–24) units of length, area, volume

g1 (Gr 1–20) units of length

m1 (Mr 1–19) units of length, area, volume, liquid capacity

h1 (H → 1–19) units of length

i1 (I → 1–14) units of length, liquid capacity, weight

Two-dimensional geometry.

a5 (Av 11–12, Br 1–3) trapezoid

b2 (Br 4–8) rectangle (field)

b4 (Bv 2–7) rectangle

c3 (Cv 1–9) isosceles triangle

d1 (Dr 1–12) isosceles triangle

d3 (Dv 1–17) right-angled triangle
f5 (Fv 1–9) quadrilateral (field)
f6 (Fv 11–17) right-angled triangle
f7 (Fv 18–22) rectangle (vineyard, χωρίον ἀμπέλιον, vines)
g3 (Gv 1–9) isosceles triangle
o2 (Or 8–13) square (field, σφραγίς)
o3 (Or 14–20) trapezoid? (vacant lot, ψιλός)
o4 (Or 22–26) triangle
o5 (Ov 1–11) quadrilateral (field?)

Three-dimensional geometry.
a2 (Ar 14–20) trapezoidal solid (breach in dike, διάκοπος ἐπὶ χώματος)
a3 (Av 1–7) trapezoidal solid
b3 (Br 10–21, Bv 1) complex solid (tower, πύργος, bricks)
b5 (Bv 8–16) trapezoidal solid
c1 (Cr 1–12) quadrilateral prism (beam?, ξύλον)
c5 (Cv 13–20) rectangular solid (excavation of river, διῶρυξ ποτάμου)
d4 (Dv 19–25) conical frustrum (circular pit, ὄρυγμα στρογγύλον)
e1 (Er 1–10) conical frustrum (circular pit, ὄρυγμα στρογγύλον)
e2 (Er 11–20) conical frustrum (circular pit, ὄρυγμα στρογγύλον)
g2 (Gr 22–28) cylinder (circular *naubion*, ναύβιον στρογγύλον)
g4 (Gv 10–17) circular or rectangular solid (circular *naubion*?, [ναύβιο]ν? στρογγύλον)
n1 (Nr 1–10) conical frustrum (cistern?)
n2 (Nr 12–17) triangular solid (granary, θησαυρὸς τρίγωνος)
n3 (Nv 1–7) rectangular solid (granary, θησαυρός, artabas of grain)
n4 (Nv 8–20) complex solid (vaulted granary, θησαυρὸς καμαρωτός)
o1 (Or 1–7) complex solid (vaulted granary, θησαυρὸς καμαρωτός)
i2 (I →16–23) rectangular solid (vat, ληνός)

Distribution of shares.
c2 (Cr 14–21) shipload of wheat
m2 (Mv 1–15) distribution of artabas of wheat

Miscellaneous problems.
d2 (Dr 13–19) speeds of runners
f1 (Fr 1–5) sale of artabas of wheat
f2 (Fr 7–18) three granaries
o6 (Ov 12–19) pay
i5 (I ↓ 16–18) unidentified

Partition into unit-fractions.
a4 (Av 8–9)
c4 (Cv 11–12)
f3 (Fr 19–21)
f4 (Fr 22–23)
h2 (H → 21–22)
i4 (I ↓ 13–14)

16. Note on Editorial Procedure

Texts in this volume are presented according to the usual papyrological practices. The following signs have their usual meanings:

( )	Resolution of an abbreviation or a symbol
[ ]	Lacuna in the text
< >	Letters omitted by the scribe
{ }	Letters erroneously written by the scribe
⟦ ⟧	Letters written, then cancelled, by the scribe
αβγδε	Letters the reading of which is uncertain or would be uncertain outside the context
. . . . .	Letters remaining in part or in whole which have not been read
[± 5]	Approximate number of letters lost in a lacuna and not restored
` ´	Letters inserted above the line by the scribe

Italicized letters on E verso were read from the photograph referred to above, Introduction section 1 at note 2. In general, abbreviations are resolved in the text. Fractions printed are d = ¼ and S = ½. The line notes correct non-standard Greek. Errors of case and varieties of spelling are very numerous.

II. Text and Translation

A recto

    - - - - -

1   [*c. 9*] .[. . .] . [.] . [

2   [*c. 8*] χαίρειν. ὁμολογῶ ἐσχηκέναι π[αρὰ σοῦ]

3   [ἐν χρήσει ἐ]ξ ὔκο σου χρ᾽υσ᾽οῦ νομισμάτιαν ε[ὐχάρα-]

4   [κτον δί]ζωτον ἓν κεφαλέου ἐπὶ δῷ μ᾽ ἀγ[τὶ]

5   [*c. 9*] καὶ τὸν εἰς λόγον ἀποτάκτου ἐπιγερδί[ας]

6   [*c. 9*]ς μηνὸς Θὼθ τοῦ ἐνεσ[τῶτ]ος ἔτ .[

7   [*c. 9*] .αθέντα ἀκυλάγ[τ]ος ἀρ[γ]υρίου τάλαντα .[

8   [*c. 8* ὅ]περ κεφάλεον [ἀ]ποδώσο σοι ὅποταν ἐρῆι

9   [ἄνευ ὑπερ]θέσεως ἢ κ[α]ὶ [εὑρ]ησιλογίας, γινομένης

10   συ τῆς πρ[άξ]εως παρά τε ἐμοῦ καὶ ἐκ τῶν ὑπαρχόντων

11   μοι πάντων. κύριων τὸ γραμμάτιων ἁπλοῦν γραφα[ὶν]

12   καὶ ἐπερωτηθεὶς ὁμολόγησα.

13   *decorative border*

14   διάκοπό(ς) τις ἐπὶ χώματος, οὗ τὸ μὲν μῆκος

15   πηχῶν λ̄, τὸ δὲ κάτω τῶ χώματο(ς) σαναωσ .[   ]

16   πηχῶν ῑ, βάθος πηχῶν ε̄, πλάτος πηχ[ῶν ϛ.]

17   οὕτω ποιοῦμαι. σηντίθω τὸ πλάτος καὶ τὸ σα .[,]

18   ϛ κα[ὶ] δέκα. γί(νεται) ῑϛ. ὧν ἥμισυ η̄. ἐπὶ τὸ βάθος, [πη-]

19   χῶν ε̄. γί(νεται) μ̄. ἐπὶ τὸ μῆκος, πηχῶν λ̄. γί(νεται) ˏΑσ̄.

20   τὰ ναύβια, παρὰ τὸν κ̄ζ̄. γί(νεται) μ̄δ γ΄ θ′′. ὅτως ἔχε[ι.]

21   *decorative border, ankh, palm fronds*

3 *l.* οἴκου | νομισματϊαν *pap.*, *l.* νομισμάτιον | 4 *l.* δίζωδον | *l.* κεφαλαίου | *l.* τῷ | 5 *l.* ἐπικερδίας | 7] .αθέντα: θ *ex* λ | *l.* ἀκοιλάντως | 8 *l.* κεφάλαιον | *l.* ἀποδώσω | *l.* αἱρῆι | 10 *l.* σοι | 11 *l.* κύριον | *l.* γραμμάτιον | *l.* γραφὲν | 12 *l.* ὡμολόγησα | 15 τῶ: *l.* τοῦ | 16 ῑ *pap.* | πηχῶν²: χ *corr. ex* ω | πηχῶν³: π *corr. e* χ (?) | 17 *l.* συντίθω | 18 δ̄εκα *pap.* | ἥμισυ: υ *corr. ex* η | 20 γθ′′ *pap.* | *l.* οὕτως

A recto

(a1: model contract)

- - - - -

1	[ ].[. . .].[.].[
2	[ca 8 g]reetings. I acknowledge that I have received f[rom you]
3	[on loan f]rom your household one well-stamped, two-imaged solidus of gold [ ]
4	[ca 9]. . . in principal amount on condition that I shall . . . [
5	[ca 9] and the [ ] on account of fixed profit [
6	[ca 9 from] the [current?] month of Thoth of the present year [- -
7	[ca 9] (fixed?) talents of money in full [
8	[ca 8] which principal amount I shall repay to you whenever you choose
9	[without post]ponement or excuse, with you having the (right of)
10	execution from me and all my possessions.
11	The agreement, written in one copy, is authoritative,
12	and on having been asked the formal question, I assented.
13	*decorative border*

(a2: mathematical problem)

14	A certain breach in a dike, whose length (is)
15	30 cubits, the lower (dimension) of the dike . . .
16	10 cubits, depth 5 cubits, breadth [6] cubits.
17	I proceed as follows. I add the breadth and the . . .
18	6 and ten. The result is 16. Half of this is 8. Times the depth,
19	5 cubits. The result is 40. Times the length, 30 cubits. The result is 1200.
20	(To get) the naubia, (I divide) by 27. The result is 44⅓ 1/9. This way (for similar cases).
21	*decorative border, ankh, palm fronds*

A verso

- - - - -

1   [. . . . .] πηχῶν ⸢ϛ⸣, πλάτος τακ[τύλων η̄,]

2   [πάχος δ]ακτήλων δ̄. οὕτω ποιοῦμ[εν. συντίθω]

3   [η̄ καὶ ϛ̄.] γί(νεται) ῑδ̄. ὧν ἥμισυ ζ̄. ἐπὶ τὸ πλ[άτος, ζ̄ ἐπὶ]

4   [η̄.] γί(νεται) ν̄ϛ̄. ἐπὶ δὸ πάχος, τακτήλ[ων δ̄. γί(νεται) σκ̄δ̄.]

5   [παρὰ] τὸν σπ̄η̄, καὶ τὰ ληπὰ εἰς [δακτύλους.]

6   [γί(νεται)] τάκτη[λ]οι ῑη̄ ῶ̄. οὕτως ἔχει ὁμ[οίως.]

7   *diagram* [π]εριφέρια πηχ()″ | πλάτος δακ[   ] | πάχος δα[   ] | ιη

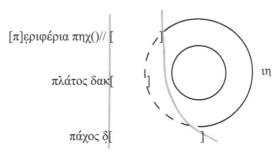

[π]εριφέρια πηχ()″ [   ]

πλάτος δακ[ιη

πάχος δ[   ]

8   λοιπαὶ δ̄, τὸ π̄ ἐν δ̄ μόρια· μὴ πρόβα ρ̄.

9   ἔσται τὸ μόρια ο, οη, πδ, ρε. ″

10   *decorative border*

11   τραπάρδιον ἰσοσκελὲς εἶ τὰ σκέλι ἀνὰ σ-

12   χοινία ῑε̄, κοινὴ βάσις λ̄, κορυφῆς ϛ̄. ἐπει-

13   δὴ ἀπὸ τῆς κηνῆς βάσεως ἐφήλω τὴν κορη-

14   φὴν, ἀπὸ τῶν λ̄ οἰφέλωμεν ϛ̄. λοιπὲ κ̄δ̄.

15   ὧν ἥμισυ ῑβ̄. ἔσται ἡ βάσις τῶ ὀρθογωνίου.

16   [τὰ ῑ]ε̄ τοῦ ἑκάστου ὀρθογωνίου ἐφ' ἑαυτά.

17   [γί(νεται)] σ̄κ̄ε̄. καὶ τὰ ῑβ̄ ἐφ' ἑαυτά. γί(νεται) ρ̄μ̄δ̄. λοιπὲ π̄ᾱ.

18   ὧν πλευρὰ θ̄. ἔσται ἡ βάσις τοῦ τετραγώνου θ̄.

19   εὑρεῖν δὸ ἐνβαδόν. ῑβ̄ ἐπὶ δὸν θ̄. γί(νεται) ρ̄η̄.

20   ὧν ἥμισυ ν̄δ̄. εὑρεῖν τὸ ἐνβαδὸν τοῦ τετρα-

21   [ο]ρθογωνίου. θ̄ ἐπὶ τὸν ϛ̄. γί(νεται) ν̄δ̄. ἔσται

1 *l.* δακτύλων | 2 *l.* δακτύλων | 4 *l.* τὸ | *l.* δακτύλων | 5 ληπὰ: π *corr. ex* γί(νεται), *l.* λοιπὰ | 6 *l.* δάκτυλοι | 7 *l.* περιφέρεια | 8 *l.* π′ | 9 *l.* τὰ | *l.* ο′, οη′, πδ′, ρα′ | 11 *l.* τραπέζιον (?) | ἰσοσκελες *pap.* | εἶ: *l.* οὗ (?) | σκέλι: κ *corr.*, *l.* σκέλη | 12 *l.* κορυφὴ | 13 *l.* κοινῆς | *l.* ὑφέλω | 13–14 *l.* κορυφὴν | 14 *l.* ὑφέλομεν | *l.* λοιπαὶ | 15 *l.* τοῦ | 17 *l.* λοιπαὶ | 19 *l.* τὸ ἐμβαδόν | *l.* τὸν | 20 *l.* ἐμβαδὸν

A verso

(a3: mathematical problem)

- - - - -

1   … 6 cubits, breadth [8] fingers,
2   [thickness] 4 fingers. We proceed as follows. I add
3   [8 and 6.] The result is 14. Half of this is 7. Times the breadth, [7 times]
4   [8.] The result is 56. Times the thickness, [4] fingers. [The result is 224.]
5   [(I divide) by] 288, and the remainder (converted) into [fingers].
6   [The result is] 18⅔ fingers. This way for similar cases.
7   *diagram*

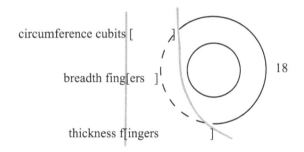

circumference cubits [   ]

breadth fing[ers   ]

18

thickness f[ingers   ]

(a4: mathematical problem)
8   Remainder 4, 1/80 (of it) in 4 unit-fractions; do not surpass 100.
9   The unit-fractions will be 1/70, 1/78, 1/84, 1/95 (*error for* 1/91).
10   *decorative border*

(a5: mathematical problem)
11   An isosceles trapezoid? whose legs are each
12   15 schoinia, common base 30, top (dimension) 6. Since
13   from the common base I subtract the top (dimension),
14   we subtract 6 from 30. The remainder is 24.
15   Half of this is 12. This will be the base of the right-angled (triangle).
16   [The] 15 of each right-angled (triangle) times itself.
17   [The result is] 225. And the 12 times itself. The result is 144. The remainder is 81.
18   The square root of this is 9. The base of the rectangle will be 9.
19   To find the area. 12 times 9. The result is 108.
20   Half of this is 54. To find the area of the rectangle.
21   9 times 6. The result is 54.

B recto

1　　[τὸ ἐμβαδ]ὸν τοῦ τετραγών[ου ν̅δ̅. οὕτως]
2　　[ἔχει ὁμο]ίως.//
3　　*diagram* ιε | ν̣δ | θ | νδ | σ̅κ̅ε̅ | ιε | ιβ | ρμδ

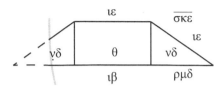

4　　[χωρ]ίον τετράγωνον εἰς ὃ πλ[ευρὸν ἔχον]
5　　[ἀπὸ ν]ότου εἰς βορρᾶς σχοινίων ὅσαδί[ποτε,]
6　　[ἀπὸ λι]βὸς εἰς ἀπηλιώτου σχοινίων δ̅ [ ]
7　　[οὕτω π]οιοῦμαι. τὰ δ̅ ἐφ' ἑαυτά. γί(νεται) ι̅ϛ̅. λεγ[ ]
8　　αὐτὸ τὸ ἐ̣ν̣βατὸν ⌜ι̅ϛ̅⌝. οὕτως ἔχει.
9　　*decorative border*

10　　πόργος κρηπίδας ἔχων περὶ τῆς κρηπ[ῖδος　]
11　　ἔχων ποιχῶν κ̅, ἡ δὲ ἐσοτέρα πηχῶν ι̣η̣, [τὸ πάχος]
12　　πηχῶν β̅, ὕψος πηχῶν ξ̅. τὸ τὲ μῆκος
13　　τοῦ τετραγώνου ῐ, τ[ὸ] πλάτος πηχῶν η̅. εὑρε[ῖν]
14　　πόσος πλίνθους δ̣ ̣απανηαι. οὕτω πο[ι-]
15　　οῦμεν. συντίθω κ̅ κ̣α̣ὶ̣ ιη. γί(νεται) λ̅η̅. ὧν ἥμ̣[ισυ, ι̅θ̅.]
16　　ἐπὶ τὰς τοῦ πάχους β̅. γί(νεται) λ̅η̅. ἐπὶ τὸ ὕψος ⌜λ̅[η̅]⌝ ἐπὶ τὸν]
17　　ξ̅. γί(νεται) Ϗσπ. ὁμοίως καὶ τοῦ τετραγώνου [μῆκος ῐ]
18　　ἐπὶ τὸν ξ̅. γί(νεται) κ̅. καὶ η̅ ἐπὶ τὸν β̅. γί(νεται) ι̅ϛ̅. σϋτίθ[ω]
19　　κ̅// καὶ ι̅ϛ̅ γί(νεται) λ̅ϛ̅. [ἐπὶ] τὸν ξ̅. γί(νεται) Ϗρ̅ξ̅. συ[ντίθω]
20　　Ϗρ̅ξ̅ καὶ Ϗσπ. γί(νεται) Δυμ̅. ἐπὶ τὸν μ̅η̅. γί(νεται) [(μυριάδες) κ̅α̅]
21　　Ϗρκ α[. .] ἄρα χωρῆ [ὁ] π⟨ό⟩ργο`υ´ς πλίνθ[ους (μυριάδες) κ̅α̅ Ϗρκ.]

5 *l.* βορρᾶν | *l.* ὁσαδήποτε | 6 *l.* ἀπηλιώτην | 8 *l.* ἐμβαδὸν | 10 *l.* πύργος | κρηπίδας: η *corr. ex* ει | 11 ποιχῶν: *l.* πηχῶν | *l.* ἐσωτέρα | ι̅η̅ *corr. ex* η̅? | 12 τὲ: *l.* δὲ | 14 *l.* πόσους | 16 ὕψος: υ *corr.* | 18 ξ̅: *error for* β̅ | *l.* συντίθω | 21 *l.* χωρεῖ | π⟨ό⟩ργο`υ´ς *corr. ex* πρκος, ς *rewritten, l.* πύργος

B recto

(b1 = a5 continued)

1   The area of the rectangle will be [54.] [This way]

2   for similar cases.

3   *diagram*

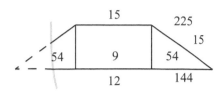

(b2: mathematical problem)

4   A rectangular field having its side

5   [from] south to north as many as?

6   [from] west to east, 4 schoinia.

7   I proceed [as follows.] The 4 times itself. The result is 16. I say? . . .

8   the area is 16. This way (for similar cases.)

9   *decorative border*

(b3: mathematical problem)

10   A tower having substructures?, having around the substructure? . . .

11   20 cubits, the inner (perimeter) 18 cubits, [the thickness]

12   2 cubits, height 60 cubits. The length

13   of the rectangle 10 (cubits), breadth 8 cubits. To find

14   how many bricks .... We proceed as follows.

15   I add 20 and 18. The result is 38. Half of this is [19].

16   Times the 2 of the thickness. The result is 38. Times the height, 3[8 times]

17   60. The result is 2280. Likewise also of the rectangle, [the length, 10,]

18   times 60 (*error for* 2). The result is 20. And 8 times 2. The result is 16. I add

19   20 and 16. The result is 36. Times 60. The result is 2160. I add

20   2160 and 2280. The result is 4440. Times 48. The result is [21]3,120

21   . . . Hence the tower contains [213,120] bricks.

B verso

- - - - -

1   *diagram* ξ | ν̣δ | η | [   ] .

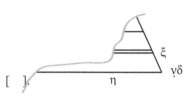

2   [τετράγ]ωνον εἰς ὃ πλευρὸν ἔχο[ν ἀπὸ νότου]
3   [εἰς βορρ]ᾶς σχοινίων λ̄, καὶ βάσις [σχοινί-]
4   [ων ῑγ. ε]ὑρεῖν τὸ ἐνβαδόν. οὕτω μ[τὸ λ̄]
5   [ἐφ’ ἑα]υτά. γί(νεται) λ̄. ἐπὶ τὸν ῑγ. γί(νεται) (μυριὰς) ᾱ [ʽΑψ. παρὰ]
6   [τὸν ⸢λ̄⸣]. γί(νεται) τ̄ρ. οὕτως ἔχει ὁμοίως.
7   *diagram* λ | τρ | τρ | ιγ | (μυριὰς) α ʽΑψ

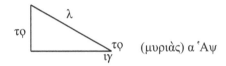

8   [τρα]πέσδιον τετράγωνον, τὸ μὲν μῆκος πηχῶν
9   [μ̄η,] τὸ τὲ πλάτος πηχῶν ι, πάχος τακτύλων ε̄,
10   [κο]ρηφὴ τακτήλων β̄. οὕτω ποιοῦμεν· συν-
11   [τίθ]ω τὸ πλάτος καὶ τὴν κορηφύν, ῑ καὶ β̄. γί(νεται) ῑβ.
12   [ὧ]ν ἥμησυ, ϛ̄. πάλιν πολυπλαδιάσζωμεν ἐπὶ
13   [τ]ὰς τοῦ πάχους ε. γί(νεται) λ̄. ἐπὶ τὸ μῆκος, πηχῶν
14   [⸢μ̄η⸣.] γί(νεται) ʽΑυμ. παρὰ τὸν ρ̄ϙβ, καὶ τὰ ληπὰ ἰς τακτύλος.
15   [γί(νεται)] ζ̄ καὶ δακτύλου ῑβ. οὕτως ἔχει ὁμοίως.
16   *diagram* | μῆκος πηχ | μη | ιβ | ρϙβ | ζ καὶ δακτήλων ῑβ | οὕτως ἔχει

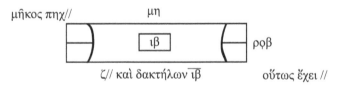

17   *palm frond and ankh*

3 *l.* βορρᾶν | 4 τὸ *corr. ex* δὸ | *l.* ἐμβαδόν | 8 *l.* τραπέζιον | 9 τὲ: *l.* δὲ | ϊ *pap.* | *l.* δακτύλων | 10 *l.* κορυφὴ | *l.* δακτύλων | 11 *l.* κορυφήν | 12 *l.* ἥμισυ | *l.* πολυπλασιάζομεν | 14 *l.* λοιπὰ εἰς δακτύλους | 15 *l.* δάκτυλοι | 16 *l.* δακτύλων

B verso

1   *diagram*

(b4: mathematical problem)

2   A rectangle having its side [from south]

3   [to] north 30 schoinia, and base [13 schoinia.]

4   To find the area. I? proceed? as follows. ... [The 30]

5   [times] itself is 900. Times 13. The result is 1[1,700. (I divide) by]

6   [the 30.] The result is 390. This way for similar cases.

7   *diagram*

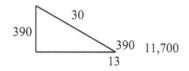

(b5: mathematical problem)

8   A quadrangular trapezoid having length [48] cubits,

9   breadth 10 cubits, thickness 5 fingers,

10   top (dimension) 2 fingers. We proceed as follows. I add

11   the breadth and the top (dimension), 10 and 2. The result is 12.

12   Half of this is 6. Again we multiply by

13   the 5 of the thickness. The result is 30. Times the length, [48] cubits.

14   The result is 1440. (I divide) by 192, and (convert) the remainder into fingers.

15   [The result is] 7 and 12 fingers. This way for similar cases.

16   *diagram*

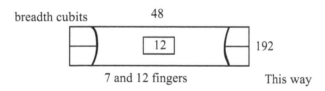

17   *ankh*

C recto

1   ξύλων νέον πηχ[ῶν κ̅η̅   ]

2   . .[. .] ρ̣ι ̣οι δακτύλων ι̅ς̅, τ̣ὸ δὲ̣ .[ ]

3   [δα]κτύλων ι̅β̅, πάχος μὲν α̣π̣ο̣ρ̣ι δακτύ[λων]

4   [η̅ . . . . .]φύλλων δακτύλων ς̅. εὑρ[εῖν]

5   [6–7] τ̣ὸ ξύλον. οὕτω ποιοῦμαι. συντί̣[θω]

6   [τὸ πλ]άτος, ι̅ς̅ καὶ ι̅β̅. γί(νεται) κ̅η̅. ὧν ἥμισυ, [ι̅δ̅. συν-]

7   [τί]θο̣μεν τὸ πάχος, {ι}ς̅ καὶ η̅. γί(νεται) ι̅δ̅. ὧν ἥμι[συ, ζ̅.]

8   [ἐ]π̣ὶ δὲ τ[ὸ]ν ι̅δ̅. γί(νεται) ο̅η̅. ἐπὶ τὸ μῆκος, πηχῶν κ̅η̅.

9   [γί(νεται)] Ϛ̅Β̅ψμδ. τ̣αῦτα μερίζομαι π[α]ρ̣[ὰ] τὸν σ̅π̅η̅, [ ]

10   καὶ τὰ λυπὰ̣ ι̣ς δακτύλους {ι̅β̅}. γί(νεται) θ̅ καὶ δακτύλ[ων]

11   ι̅β̅ ω̅. οὕτως ἔχει ὁμοίως.⫽

12   *diagram* ς̣ | κη | ι̅ς̣ | ι̅β

13   *decorative border*

14   ἀναίβαλέν τις εἰς πλοῖον ἀπ[ὸ] τοῦ θυσαυροῦ τ[ὸ]

15   ἥμισ̣η,′ καὶ ἰς δὸ τημόσιον τὸ [τ]ρ̣ί[το]ν̣ καὶ ὑπὲρ μισθ[οῦ]

16   ὀνηλάτου τὸ δωδέκατον, κ[αὶ] κ̣α̣τ̣α̣λίφθης ἰς πλοῖ-

17   ον ποιροῦ ἀ(ρτάβα)ς ν̅. οὕτω ποιοῦμε̣ν. συντίθω τὸ ἥμισ-

18   υ καὶ τὸ τρίτον καὶ τὸ δωδέκατ[ο]ν. γί(νεται) [Ϛ̅]′ ⟨γ̅⟩ ι̅β̅. τί λίπιν

19   δὺν μονάτον μίαν; ι̅β̅. ι̅β̅ ἐπὶ τ[ὸ]ν ν̅. γί(νεται) χ̅.

20   ἄρα χωρήσι τὸ πλοῖων ποιρ[ο]ῦ̣ ἀ(ρτάβα)ς χ̅. οὕ[τ]ω̣[ς ἔ]χει

21   ὁμοίως.⫽⫽⫽

---

1 *l.* ξύλον | 3 δακτυ: δ *corr. ex* γ? | 10 *l.* λοιπὰ εἰς | 14 *l.* ἀνέβαλέν | θυσαυροῦ: υ² *corr. ex* ρ, *l.* θησαυροῦ | 15 *l.* ἥμισυ | ἴς *pap.* | *l.* εἰς τὸ δημόσιον | ὕπερ *pap.* | 16 *l.* κατελείφθησαν | ἴς *pap.*, *l.* εἰς | 17 *l.* πυροῦ ἀ(ρτάβαι) | 18–19 *l.* τί λείπειν τὴν μονάδα μίαν (*scil.* τί λείπει ἡ μονὰς μία) | 20 *l.* χωρήσει | *l.* πλοῖον | *l.* πυροῦ

C recto

(c1: mathematical problem)

1   A fresh beam?, [28] cubits . . .

2   . . . 16 fingers, and the . . .

3   12 fingers, thickness . . . [8] fingers

4   . . . leaves? 6 fingers. To find

5   the beam?. I proceed as follows. I add the

6   breadth, 16 and 12. The result is 28. Half of this is [14] We add

7   the thickness, 6 and 8. The result is 14. Half of this is [7.]

8   [7] times 14. The result is 98. Times the length, 28 cubits.

9   [The result is 2]744. This I divide by 288, . . .

10   and (I convert) the remainder into fingers. The result is 9 and 12⅔ fingers.

11   This way for similar cases.

12   *diagram*

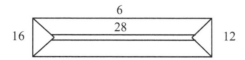

13   *decorative border*

(c2: mathematical problem)

14   Someone loaded on a boat, from the granary

15   half, and for the taxes one third, and for the pay

16   of the donkey-driver one twelfth, and there remained on the boat

17   50 artabas of wheat. I proceed as follows. I add one half

18   and one third and one twelfth. The result is [½] <⅓> 1/12. How much

19   is a unit (i.e. 1) lacking? 1/12. 12 times 50. The result is 600.

20   Hence the boat will hold 600 artabas of wheat. This way

21   for similar cases.

C verso

1   [τρίγωνον ἰσο]σκελὶ ἀνὰ σχοινία ͞ιη, κε̣ν̣ὴ [βά-]
2   [σις ͞μη. ε]ὑρεῖν τὰς ἄλλας πλευράς. (*vac.*)
3   [σχο]ι̣νία ͞ιη ἐφ' ἑαυτά. γί(νεται) ⟨τ⟩κδ. καὶ τὸ ἥμη[συ]
4   [τ]ῆς βάσις, ͞κδ. ͞κδ ἐφ' αὑτά. γί(νεται) φ[ο̅ϛ̅. φο̅ϛ̅ καὶ]
5   [τ]͞κδ. γί(νεται) ͞λ. ὧν πλευρὰ ͞λ. ἄρα ἦν [ἡ πλευρὰ]
6   [͞λ]. εὑρεῖν καὶ τὸ ἐβαδόν. οὕτο ποιοῦμ̣[εν.]
7   [τ]ὴν βάσιν ἐπὶ τὴν ἑκάστην ὀρθήν, ͞μ[͞η ἐπὶ]
8   [τὸ]ν ͞λ. γί(νεται) Ἀυμ. ὧν ἥμισυ ψ̅κ̅. οὕτως ἔ[χει]
9   [ὁ]μοίως.͞//
10 *diagram* λ | ιη | ψκ | μη

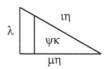

11 λοιπαὶ α d, τὸ ͞ζ ἐν ͞ζ μόρια· μὴ ⟨πρόβα⟩ ͞ρ. ἔσται τὰ
12 μόρια [*vac.?*] κδ, κη, λε, μβ, νϛ, ξ, ͞ο.

13 διῶραξ π̣ο̣ταμοῦ, μῆκο⟨ς⟩ σχοινία ͞β d, πλάτος
14 ξύλων ͞λ, βάθο[ς] πηχῶν ͞ε. οὕτο ποιοῦμεν.
15 ἔχει τὸ σχ̣οινίω[ν] πήχεις ͞ο̅ϛ̅. β d ἐπὶ τὸν ͞ο̅ϛ̅.
16 γί(νεται) ͞σι̅ϛ̅. καὶ ἔχει τ[ὸ] ξύλων πηχῶν ͞γ. ͞γ ἐπὶ τὸν
17 ͞λ. γί(νεται) ͞ο̣. ο// ἐπὶ τὸν ͞σι̅ϛ̅. γί(νεται) (μυριὰς) α Ὀυμ. ἐπ[ὶ] τὸ βά-
18 θος, πηχῶν ͞ε. γί(νεται) (μυριάδες) θ Ζσ. ἔστι ναύβια, παρὰ
19 τὸν ͞κζ. γί(νεται) Ͳγχ. οὕτως ἔχει ὁμοίως.
20 *diagram* σχοι͵ β (d) | ξ[υ]λ̣ ͞λ | . . | Ͳγχ

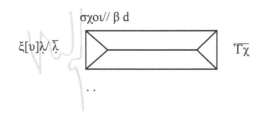

1 *l.* ἰσοσκελὲς | *l.* κοινὴ | 3 ͞ιη *pap.* | *l.* ἥμισυ | 4 *l.* βάσεως | 6 *l.* ἐμβαδόν | *l.* οὕτω | 11 *l.* ͞α | ξ͵: *l.* ζ͵ | ͞μη̅ρ̅ *pap.* | *l.* κδ', κη', λε', μβ', νϛ', ξ', ο' | 13 *l.* διῶρυξ | μῆκος: κ *corr.* | 14 *l.* οὕτω | 15 *l.* σχοινίον | 16 *l.* ξύλον

C verso

(c3: mathematical problem)

1   An isosceles [triangle] with (legs) 18 schoinia each common [base]

2   [48.] To find the other sides.

3   18 schoinia times itself. The result is ⟨3⟩24. And half

4   of the base is 24. 24 times itself. The result is 5[76. (I add) 576 and]

5   [3]24. The result is 900. The square root of this is 30. Hence [the side] was

6   [30.] To find also the area. We? proceed as follows.

7   The base times each vertical, 4[8 times]

8   30. The result is 1440. Half of this is 720. This way

9   for similar cases.

10   *diagram*

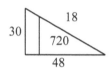

(c4: mathematical problem)

11   Remainder 1 ¼, 1/7 (of it) in 7 unit-fractions; do not surpass 100.

12   The unit-fractions will be 1/24, 1/28, 1/35, 1/42, 1/56, 1/60, 1/70.

(c5: mathematical problem)

13   A trench of a canal, having length 2 ¼ schoinia, breadth

14   30 xyla, depth 5 cubits. We proceed as follows.

15   One schoinion contains 96 cubits. 2 ¼ times 96.

16   The result is 216. And one xylon contains 3 cubits. 3 times

17   30. The result is 90. 90 times 216. The result is 19,440. Times the depth,

18   5 cubits. The result is 97,200. (I convert?) into? naubia, (I divide) by

19   27. The result is 3600. This way for similar cases.

20   *diagram*

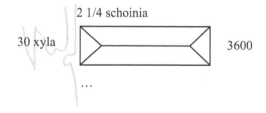

D recto

- - - - -

1   [6–7] ̣ ̣αδυο ̣ [c. 17]
2   [. . . .] ̣[̣] ἰ[σ]ο̣σκελὴς ἀνὰ σχοινία [κ̄ c. 8]
3   [κ]εν[ὴ] βάσις λ̄β̄. εὑρεῖν τὰς ἄλλας πλευ[ράς. οὕ-]
4   [τω] ποιοῦμαι. λαβάνωμεν τὸ ἥμυσυ [τῆς βά-]
5   [σε]ω̣ς, ῑϛ̄. ἐφ' ἑαυτά. γί(νεται) σ̄ν̄ϛ̄. καὶ τὰ κ̄ ἐφ' [ἑαυτά.]
6   [γί(νεται)] ῡ. ἀπὸ τῶ ῡ ὑφέλομεν σ̄ν̄ϛ̄. λοιπαὶ ρ̄μ̄[δ̄.]
7   [ὦν πλ]ευρὰ ῑβ̄. ἄρα ἦν ἡ ἑκάστη ὀρθὴ ῑβ̄. εὑ[ρεῖν]
8   [τὸ ἐ]μβαδόν. τὴν βάσιν ἐπὶ τὴν ὀρθήν, ιβ [ἐπὶ]
9   [τὸ]ν̣ ῑϛ̄. γί(νεται) ρ̄ϙ̄β̄. ὦν ἥμισυ ϙ̄ϛ̄. ἄρα ἦν τὸ [ἐμ-]
10   β[α]δὸν ἀρουρῶν ϙ̄ϛ̄. οὕτως ἔχει ὁ[μοί-]
11   ως.///////
12   *diagram* ἄρα ἦ̣ν αὐτὴ ῑβ̄ | ρ̄ϙ̄β̄ | ῑϛ̄ | ϙ̄ϛ̄ | κ̣ι̣νὴ βάσις λ̄β̄ | ῑϛ̄

ἄρα ἦν αὐτὴ ῑβ̄   ρ̄ϙ̄β̄
ῑϛ̄   ϙ̄ϛ̄
κινὴ βάσις λ̄β̄ ῑϛ̄

13   ἔ̄τρεχέν τις εἰς λ̄β̄ ἡμέρας σδάδ[ι]α θ̄. μηθ' ἡμ[έρας]
14   ῑβ̄ ἕτερος ἐπ[αν]ελθὼν ἔτρεχεν σ̣[τ]ά̣δια [ῑ]ε̄.
15   εὑρεῖν ἐν πόσ[αις] ἡμέρας ὁ δεύτερος ἀ[ν]ταλή[ψεται]
16   τὸν πρῶτο̣ν̣. ο̣ὕτω ποιοῦμεν. ῑβ̄ ἐπὶ τὸν [θ̄.]
17   γί(νεται) ρ̄η̄. ἀπὸ τῶ̣ν [ῑ]ε̄ ὑφέλομα̣ι θ̄. λοιπαὶ ϛ̄.
18   τὸ ρη τὸ ὠγδοωκετέκατον, ϛ̄. ἄρα ὁ τεύτε[ρος]
19   [ἀ]ν̣δαλύψητα̣[ι] [τ]ὸν πρῶτον ἐν ῑη̄ ἡμέρας.

3 *l.* κοινὴ | 4 *l.* λαμβάνομεν *l.* ἥμισυ | 6 *l.* τῶν |12 ϙϛ: ϛ *corr. ex* c | *l.* κοινὴ | 13 *l.* στάδια | *l.* μεθ' | 15 *l.* ἡμέραις | *l.* ἀντιλήψεται | 18 τὸ¹: *l.* τοῦ | *l.* ὀκτωκαιδέκατον | *l.* δεύτερος | 19 *l.* ἀντιλήψεται | *l.* ἡμέραις

D recto

(d1: mathematical problem)

- - - - -

1   . . . two . . .
2   . . . isosceles [triangle?] with (legs) [20] schoinia each . . .
3   common base 32. To find the other sides. I proceed thus.
4   We take half of the base,
5   16. Times itself. The result is 256. And the 20 times [itself.]
6   [The result is] 400. We subtract 256 from 400. The remainder is 14[4].
7   The square root of this is 12. Hence each vertical was 12. To find
8   the area. The base times the vertical, 12 [times]
9   16. The result is 192. Half of this is 96. Hence the area was
10   96 arourai. This way for similar
11   cases.
12   *diagram*

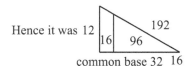

Hence it was 12 | 192 | 16 | 96 | common base 32 16

(d2: mathematical problem)
13   A person ran for 32 [*error for* 12] days, 9 stades (per day). After
14   12 days another person coming after ran [1]5 stades (per day).
15   To find in how many days the second person will catch up with
16   the first. We proceed as follows. 12 times [9].
17   The result is 108. I subtract 9 from 15. The remainder is 6.
18   One eighteenth of 108 is 6. Hence the second person
19   will catch up with the first in 18 days.

D verso

- - - - -

1   [c. 17] . . . . . [c. 11]
2   [c. 9] ε[ὑ]ρ[ε]ῖν τ[ὰς] ἄλλας πλευράς. οὕτω π[ο]ι[οῦμεν.]
3   [5-6] πεθακωρικὸν ὀρθογώνιον γ̅δ̅ε̅, [ὀρθὴ γ̅, βά-]
4   [σις] δ̅, ὑποτίνουσα ε̅, σὺν περιοχῇ ι̅β̅. συντ[ίθω τὰ]
5   [γ̅′] δ̅″ ε̅″ καὶ ι̅β̅. γί(νεται) κ̅δ̅. ἀπόδιξει[ς]. μερίζο[μαι τὸν]
6   [ρ̅ϟ̅β̅] παρὰ τὸν κ̅δ̅. γί(νεται) η̅″. ἐπὶ τὸν γ̅. διὰ τί ἐπ[ὶ γ̅; ὅτι]
7   [ἡ] ὀρθὴ γ̅. η̅″ ἐπὶ τὸν γ̅. ⟨γί(νεται)⟩ κ̅δ̅″. ἔσται ὀρθ[ὴ κ̅δ̅.]
8   [καὶ] η̅ ἐπὶ τὸν δ̅. γί(νεται) λ̅β̅. καὶ η̅ ἐπὶ τὸν ε̅. γί(νεται) μ̅″. [c. 4] ἄρα ἦ[ν]
9   [ἡ] βάσις λ̅β̅, ὑποτίνουσα μ̅. εὑρεῖν τὸ ἐνβα[δόν]. τὴν
10   [βά]σιν ἐπὶ τὴν ὀρθήν, κ̅δ̅ ἐπὶ λ̅β̅. γί(νεται) ψ̅ξ̅η̅. ὧν ἥμισυ
11   [τ]π̅δ̅. ἔσται τὸ ἐνβαδὸν ἀρουρῶν τ̅π̅δ̅. εὑρεῖν καὶ τὴν
12   [π]εριοχήν. συντίθω κ̅δ̅ καὶ λ̅β̅ καὶ μ̅. γί(νεται) ϟ̅ϛ̅. ἔσται τὴν
13   [π]εριοχὴν ϟ̅ϛ̅. καὶ συντίθω τὴν ὀρθὴν καὶ τὴν βάσιν
14   [κ]αὶ τὴν ὑποτίνουσα, κ̅δ̅ καὶ λ̅β̅ καὶ μ̅. γί(νεται) ϟ̅ϛ̅. ἄρα ἦν ἡ ὀρθ[ὴ]
15   [σ]ὺν βάσις σὺν ὑποτινούσᾳ ϟ̅ϛ̅. καὶ σὺν περιοχῇ, ϟ̅ϛ̅.
16   [σ]υντίθω ϟ̅ϛ̅ καὶ ϟ̅ϛ̅. γί(νεται) ρ̅ϟ̅β̅. οὕτως ἔχει ὁμοίως.″
17   decorative border, diagram κδ | ρϟβ | τπ̅δ̅ | μ, decorative border

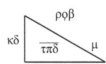

18   decorative border

19   [ὄρυ]γμα στροκύλουν, ἡ ἄνω διάμετρος π[η]χῶν
20   [ὁσ]ωνδήποτε, τὸ βάθος πηχῶν γ̅. ἐπὶ ναύβια κ̅α̅ γ̅″.
21   [οὔ]τω ποιοῦμαι. ἀναλοίω τὰ ναύβια εἰς πήχεις.
22   [ἔ]χει τὸν ναύβιον πήχεις κ̅ζ̅. κ̅α̅ γ̅″ ἐπὶ τὸν κ̅ζ̅.
23   γί(νεται) φ̅ο̅ϛ̅. παρὰ τὸν τὸ βάθος, πηχῶν γ̅. γί(νεται) ρϟ[β̅.]
24   [τ]ούτων προστίθωμεν τὸ τρίτων, ξ̅δ̅. συντίθω
25   ρ̅ϟ̅β̅ καὶ ξ̅δ̅. γί(νεται) σ̅ν̅ϛ̅. ὧν πλευρὰ ι̅ϛ̅.

3 l. πυθαγορικὸν | 4 l. ὑποτείνουσα | 5 l. ἀπόδειξις | 9 l. ὑποτείνουσα | l. ἐμβαδόν | 10 κ̅δ̅ corr.? |
11 l. ἐμβαδὸν | 12–13 l. ἡ περιοχὴ | 14 l. ὑποτείνουσαν | 15 l. βάσει | l. ὑποτεινούσῃ | 19 l.
στρογγύλον | 21 l. ἀναλύω | 22 τὸν¹: l. τὸ | 24 l. προστίθομεν | l. τρίτον

D verso

(d3: mathematical problem)

- - - - -

1 . . .

2 . . . To find the other sides. [We proceed] as follows.

3 . . . Pythagorean right-angled (triangle) 3–4–5, [vertical 3, base]

4 4, hypotenuse 5, with perimeter 12. I add

5 [3,] 4, 5, and 12. The result is 24. Demonstration. I? divide

6 [192] by 24. The result is 8. Times 3. Why by [3? Because]

7 [the] vertical is 3. 8 times 3. ⟨The result is⟩ 24. The vertical will be [24.]

8 [And] 8 times 4. The result is 32. And 8 times 5. The result is 40. [...] So

9 [the] base was 32, the hypotenuse 40. To find the area. The

10 base by the vertical, 24 times 32. The result is 768. Half of this is

11 [3]84. The area will be 384 arourai. To find also the

12 perimeter. I add 24 and 32 and 40. The result is 96. The perimeter will be

13 96. And I add the vertical and the base

14 and the hypotenuse, 24 and 32 and 40. The result is 96. Hence the vertical

15 with the base with the hypotenuse was 96. And with the perimeter, 96.

16 I add 96 and 96. The result is 192. This way for similar cases.

17 *diagram*

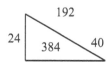

18 *decorative border*

(d4: mathematical problem)

19 A circular excavation, whose upper diameter is some number of cubits,

20 the depth 3 cubits. (Its volume) at 21⅓ naubia.

21 I proceed as follows. I convert the naubia to cubits.

22 One naubion contains 27 cubits. 21⅓ times 27.

23 The result is 576. (I divide) by the depth, 3 cubits. The result is 19[2.]

24 We add a third of this, 64. I add

25 [1]92 and 64. The result is 256. The square root of this is 16.

E recto

1   [ἄρα ἦν ἡ διά]μετρος πηχῶν ι̅ϛ̅. [   ]
2   [ὄ]ρυγμα στρονκύλουν, ἡ ἄνω διάμετρος [vac.?]
3   πηχῶν ι̅ϛ̅, τὸ βάθος πηχῶν γ̅. εὑρεῖν τ[ὰ ναύ-]
4   βια. [ο]ὕτω ποιοῦμεν. τὰ ι̅ϛ̅ τῆς διάμετρος [ἐφ’ ἑαυ-]
5   τά. γ[ί(νεται)] σ̅ν̅ϛ̅. τούτων ἡφέλωμεν τὸ τέταρτ[ον,]
6   ξ̅δ̅. ἀ̣[π]ὸ τῶν σ̅ν̅ϛ̅ οἰφέλωμεν ξ̅δ̅. λοιπὲ ⌈ρ̅[ϙ̅β̅⌉.]
7   [ἐ]π̣ὶ τὸ βάθος, πηχῶν γ̅. γί(νεται) φ̅ο̅ϛ̅. ἔστι τὰ ν[αύβια,]
8   π̣[αρὰ] τὸν κ̅ζ̅. γί(νεται) κ̅α̅ γ̅⫽. ἔσται ναύβια κ̅α̅ γ̅. οὕτω[ϛ]
9   ἔχ̣[ε]ι ὁμοίως.
10   *paragraphos, diagram* ιϛ | κα γ⫽ | γ

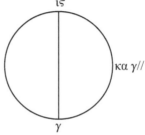

11   ὄρυγμα σ̣τρονκύλουν ἡ οὗ περ̣ειφ[έρεια]
12   πηχῶν ὁσοντήποτε, τὸ βάθ[ος π]ηχ̣[ῶν θ̅.]
13   ἀνηβλήθη ναύβια ι̅ϛ̅. εὑρεῖν τὴν π̣ερ̣[ι]πέρ-
14   ειαν. οὕτω ποιοῦμεν. ἀναλοίω τὰ ν̣άβι̣α̣
15   εἰς πήχεις. ι̅ϛ̅ ἐπὶ τὸν κ̅ζ̅. γί(νεται) υ̅λ̅β̅. παρὰ τ[ὸ]
16   βάθος, πηχῶν θ̅. γί(νεται) μ̅η̅. ἐπ[ὶ] τὸν ι̅β̅ τ[ῆς]
17   περιφερίας. γί(νεται) φ̅ο̅ϛ̅. ὧν πλευρὰ κ̅δ̅. ἄρα ἦ[ν]
18   ἡ ἄνω περιφέρηα πηχῶν κ̅δ̅. οὕτως ἔχει
19   ὁμοίως.
20   *paragraphos, partial decorative border, diagram* ′κδ | ιϛ | θ

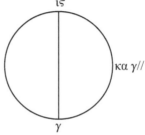

2 *l.* στρογγύλον | 4 *l.* διαμέτρου | 5 *l.* ὑφέλομεν | 6 *l.* ὑφέλομεν | *l.* λοιπαὶ | 11 *l.* στρογγύλον | *l.* περιφέρεια | 12 *l.* ὁσωνδήποτε | 13 *l.* ἀνεβλήθη | *l.* περιφέρειαν | 14 *l.* ἀναλύω | *l.* ναύβια | 15 εἰς: ι ex ϛ? | 17 *l.* περιφερείας | 18 *l.* περιφέρεια

E recto

- - - - -

1   [Hence the] diameter [was] 16 cubits. [. . .]

(e1: mathematical problem)

2   A circular excavation, whose upper diameter is

3   16 cubits, depth 3 cubits. To find the naubia.

4   We proceed as follows. The 16 of the diameter [by] itself.

5   The result is 256. From this we subtract one quarter,

6   64. We subtract 64 from 256. The remainder is 1[92.]

7   Times the depth, 3 cubits. The result is 576. The naubia:

8   (I divide) by 27. The result is 21⅓. It will be 21⅓ naubia. This way

9   for similar cases.

10   *diagram*

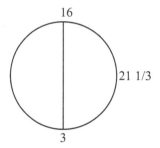

(e2: mathematical problem)

11   A circular excavation, whose circumference

12   is some number of cubits, depth [9] cubits.

13   16 naubia were excavated. To find the circumference.

14   We proceed as follows. I convert the naubia

15   into cubits. 16 times 27. The result is 432. (I divide) by the

16   depth, 9 cubits. The result is 48. Times the 12 of the

17   circumference. The result is 576. The square root of this is 24. Hence

18   the upper circumference was 24 cubits. This way

19   for similar cases.

20   *partial decorative border and diagram*

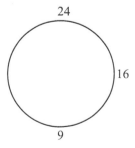

E verso

1 [ἔχει τὸ σχοινίο]ν τὸ γεωμετρικὸν ὄγ[δοα η̅, τὸ δὲ ὄγ-]
2 [δ]οον ἔχει πήχεις ι̅β̅, ὥστε εἶναι τὸ σχοι[νίον τὸ]
3 [γ]εομετρικὸν πηχῶν {ἐστιν} ο̅ς̅, τὸ εὐθημετρικόν
4 ἐστιν πηχῶν ρ̅″. ὁ εὐθημετρικὸς πῆχυς ἐστὶν
5 [ὁ] κατὰ μῆκος μόνον μετρούμενος, ἐμβαδικὸς δ[ὲ]
6 [ὁ] κατὰ μῆκο⟨ς⟩ καὶ πλάτος καὶ βάθος ἤδε ὕψος ἢ πάχος.
7 κα[ὶ] ὁ οἰκοπεδικὸς πήχεις ἔχει ἐνβατοὺς πήχεις
8 ρ̅. τὸ ξύλον ἐν ᾧ μετρεῖτε τὰ ναύβια. τὸ μὲν βασι-
9 λ[ι]κόν ἐστιν πηχῶν γ̅, παλεστῶν τὲ ι̅η̅, δ[α]κδύλων
10 ο̅β̅· τ[ὸ] δὲ ἰδιωτικόν ἐστιν πηχῶν β̅ ω̅, παλεσ-
11 τῶν δὲ ι̅ς̅, δακτύλων τὲ ξ̅δ̅, ὥστε εἶναι τὸ σχοιν⟨ί⟩ον
12 τὸ γεωμετρικὸν ξύλα βασιλικὰ μὲν λ̅β̅, ἰδιωτικά
13 δὲ λ̅ς̅. τὸ ναύβιον ἐκ τετραγώνου ἔχει ξύλον ἕν,
14 βάθος ξύλων ἕν″. τὸ ξύλων ἔχει πήχεις γ̅, ὥστε εἶναι
15 τὸ μὲν δημόσιον ναύβιον στερεῶν πηχῶν κ̅ζ̅,
16 τὸ δὲ ἰδιωτικ[ό]ν ἐστιν πηχῶν ι̅η̅ S′ ⟨γ⟩ θ νδ″. ὁ πήχεις
17 στερεὸς χωρῖ ξυροῦ ἀρτάβας γ̅ d̅ η̅, ἠγροῦ δὲ μετρη-
18 τὰς γ̅. ἡ ἄρουρά ἐστιν ἡ κατὰ πόλιν, ἡ δὲ ἐν ὀκοπέδοις
19 μετρουμένην′ βίκων ν̅. ὁ βῖκο`υ′ς ἔχει ἐνβα-
20 δοὺς πήχεις σ̅, ὥστε εἶναι τὴν ἐν οἰκοπέδοις
21 ἄρουραν ἐνβαδῶν πηχῶν μυρύων. ἡ δὲ κατ' ἄ-
22 γρον ἄρουράν′ ἐστιν βίκων μ̅η̅. ὁ βίκους ἔχει ἐνβα-
23 δοὺς πήχει λ̅ρ̅β̅, ὥστε εἶναι τὴν κατ' ἄκρον ἄρουραν
24 ἐνβαδῶν πηχῶν Ͳθσις̅. οὕτως ἔχει. *line filler*
25 *decorative border*

3 *l.* γεωμετρικὸν | *l.* εὐθυμετρικόν | 4 *l.* εὐθυμετρικὸς | πῆχυς *ex* πηχεις *(or vice versa)* | 6 *l.* ἤτοι | 7 οἰκοπεδικὸς: δ *ex* ρ | πήχεις1 *l.* πῆχυς | *l.* ἐμβαδοὺς | 8 *l.* μετρεῖται | 9 *l.* παλαιστῶν | *l.* δὲ | *l.* δακτύλων | 10 *l.* ἰδιωτικόν | 10–11 *l.* παλαιστῶν | 11 τὲ: *l.* δὲ | 12 *l.* ἰδιωτικά | 14 *l.* ξύλον (*bis*) | 15 δημόσιον: δη *ex* τυ | 16 *l.* ἰδιωτικόν | πήχεις: *l.* πῆχυς | 17 *l.* χωρεῖ | *l.* ξηροῦ | *l.* ὑγροῦ | 18 *l.* οἰκοπέδοις | 19 *l.* μετρουμένη | βίκους: *l.* βῖκος | 19–20 *l.* ἐμβαδοὺς | 21 *l.* ἐμβαδῶν | *l.* μυρίων | 22 *l.* ἄρουρά | βίκους: *l.* βῖκος | 22–23 *l.* ἐμβαδοὺς πήχεις ρ̅ρ̅β̅ | *l.* ἄγρον | 24 *l.* ἐμβαδῶν

E verso

(e3: metrological text)

1   The surveyor's schoinion [contains] [8] ogdoa, the ogdoon
2   contains 12 cubits, so that the surveyor's schoinion comprises
3   96 cubits, while the linear (schoinion)
4   comprises 100 cubits. The linear cubit is the one
5   measured only in length, and the area (cubit) is the one
6   (measured) in length and breadth and depth or height or thickness.
7   And the building-site cubit contains 100 area cubits.
8   The xylon, in which naubia are measured. The royal (xylon)
9   comprises 3 cubits, 18 palms, 72 fingers;
10   the private (xylon) comprises 2⅔ cubits, 16 palms,
11   64 fingers, so that the surveyor's schoinion
12   comprises 32 royal xyla, and 36 private (xyla).
13   The naubion contains one xylon in square,
14   depth one xylon. The xylon contains 3 cubits, so that
15   the public naubion comprises 27 solid cubits,
16   and the private (naubion) comprises 18½ ⟨⅓⟩ 1/9 1/54 cubits. The solid cubit
17   holds 3 ¼ ⅛ artabas of dry goods, and 3 metretai of fluid.
18   There is a civic aroura, and an aroura for building-sites,
19   which is measured in 50 bikoi. The bikos contains 200 area cubits,
20   so that the aroura for building-sites
21   comprises ten thousand area cubits, while the country
22   aroura comprises 48 bikoi. The bikos contains 992 (*error for* 192) area
23   cubits, so that the country aroura comprises
24   9216 area cubits. This way (for similar cases.)
25   *decorative border*

F recto

- - - - -

1   ] . ας ἀρτάβ[. .] . ποιροῦ πολῖτε (ταλάντων) . [c. 12–13]

2   πόσου. ἀπόδιξεις. η̄ ἐπὶ τὸν ῑε. γί(νεται) ρ̄κ. τόκ[ατον ῑβ.]

3   προστίθωμεν τόκατον. γί(νεται) ρ̄λ̄β. πραθήσονται [ ]

4   ἀ(ρτάβας) καὶ η̄ . με . ο ἀ(ρτάβας?) ρ̄λ̄β. οὕτως ἔχει.//////

5   *diagram* ἀ(ρτάβαι) | ιε | η η | ρ̄λ̄β

ἀ(ρτάβαι)	ιε
η η	ρ̄λ̄β

6   *decorative border*

7   θησαυρὸς τρῖς, ὁ πρῶτος θησαυρὸς ἐ̂χεν ἀ(ρτάβας) [σ̄,]

8   {ὁ τεύτερος} ὁ τεύτερος θησαυρὸς ἐ̂χεν ἀ(ρτάβας) τ̄,

9   ὁ τρίτος θησαὺ΄ρὸς ἐ̂χεν ἀ(ρτάβας) ῡ΄. εἰσῆλθόν τις

10   {τις} καὶ μίξας. εὑρεῖν ἀ(ρτάβας) χ̄λ̄. οὕτω ποιοῦμεν. σ̄ πό[σα]

11   ἑκαστοστὰς ἔχει; β̄. τριακὸς πόσα ἕκαστος ἔχε[ι; γ̄.]

12   ϋ πόσα ἑκαστηστὰς ἔχει; δ̄. συντίθω β̄ καὶ γ̄ καὶ [δ̄.]

13   γί(νεται) θ [[υ]] μερίζω χ̄λ̄ παρὰ τὸν θ̄. γί(νεται) ο̄. ο̄ ἐπὶ τὸν [β̄.]

14   γί(νεται) ρ̄μ. ὁμοίως καὶ τῶ τευτέρου, ο̄ ἐπὶ τὸν γ̄. γί(νεται) [σ̄ι.]

15   ὁμοίως καὶ τοῦ τρίτου, ο̄ ἐπὶ τὸν δ̄. γί(νεται) σ̄π. συ[ντίθω]

16   ρ̄μ καὶ σ̄ι καὶ σ̄π. γί(νεται) χ̄λ̄. οὗτος ἔχει ὁμοίω[ς.]

17   {κα}

18   *paragraphos, diagram* χλ | ρμ | σι | σπ | σ | τ | υ | χλ | οὕτως ἔχ[ει]

χλ	ρμ	σι	σπ
χλ	σ	τ	υ

οὕτως ἔχ[ει]

F recto

(f1: mathematical problem)

- - - - -

1   ... artabas of wheat remaining? talents? ...

2   of how much. Demonstration. 8 times 15. The result is 120. One tenth is [12]. ...

3   We add one tenth. The result is 132. There will be sold ...

4   artabas and 8 ... 132. This way (for similar cases).

5   *diagram*

artabas	15
8 1/8?	132

6   *decorative border*

(f2: mathematical problem)

7   Three granaries, the first granary held [200] artabas,

8   the second held 300 artabas,

9   the third granary held 400 artabas. Someone came in

10   and mixed (them) up. To find 630 artabas. We proceed as follows. 200

11   contains how many hundreds? 2. Three hundred contains how many hundreds? [3.]

12   400 contains how many hundreds? 4. I add 2 and 3 and [4.]

13   The result is 9. I divide 630 by 9. The result is 70. 70 times [2].

14   The result is 140. Likewise for the second (granary), 70 times 3. The result is [210.]

15   Likewise for the third (granary), 70 times 4. The result is 280. I add

16   140 and 210 and 280. The result is 630. This way for similar cases.

17   (canceled)

18   *diagram*

	140	210	280	
630	200	300	400	630 This way

F recto (continued)

19   λοιπὲ θ̄, τὸ ρ̅ι̅θ̅ ἐν δ̄ μόρια·

20   μὴ πρόβα ρ̄. ἔσται τὰ μόρια

21   λδ, να, ξη, πε. /////

22   λοιπαὶ λ̅ς̅ {λβ}, τὸ σ̅κ̅η̅ ἐν πέντε μόρια·

23   μοὶ πρόβα ρ̄. ἔσται τὰ μόρια ιβ, λ, νζ, ος, ϙ'ε'

24   *decorative border*

1 *l.* πυροῦ πολεῖται | 2 *l.* ἀπόδειξις | *l.* δέκατον | 3 *l.* προστίθομεν | τόκατον: τ¹ ex δ (*vel* δ ex τ), *l.* δέκατον | 7 *l.* θησαυροὶ τρεῖς | *l.* εἶχεν | 8 *l.* δεύτερος (*bis*) | τεύτερος¹: ρ *ex* υ | *l.* εἶχεν | 9 ἀ(ρτάβας): α¹ *corr. ex* ο? | *l.* εἶχεν | υ̅′ *corr.* ‖ *l.* εἰσῆλθέν | 10 *l.* μείξας | 11 ἑκαστοστὰς: *l.* ἑκατοστὰς (*melius* ἑκατὸν) | *l.* τριακόσιοι | ἕκαστος: *l.* ἑκατὸν ‖ 12 ϋ: *l.* ῡ | ἑκαστηστὰς: *l.* ἑκατοστὰς | β̅κ̅α̅ι̅γ̅ *pap.* | 13 γί(νεται) *corr. ex* καὶ | 14 *l.* τοῦ δευτέρου | 16 σ̅π̅ *corr.?* | *l.* οὕτως | 19 *l.* λοιπαὶ | *l.* ριθ′ | 21 *l.* λδ′, να′, ξη′, πδ′ | 22 *l.* σκη′ | 23 *l.* μὴ | *l.* ιβ′, λ′, νζ′, ος′, ϙε′ | ϙε *ex* ϙς

F recto (continued)

(f3: mathematical problem)

19   Remainder 9, 1/119 (of it) in 4 unit-fractions;

20   do not surpass 100. The unit-fractions will be

21   1/34, 1/51, 1/68, 1/85 (*error for* 1/84).

(f4: mathematical problem)

22   Remainder 36, 1/228 (of it) in five unit-fractions;

23   do not surpass 100. The unit-fractions will be 1/12, 1/30, 1/57, 1/76, 1/95.

24   *decorative border*

F verso

- - - - -

1   [*c.* 5 νότου β]ήματα ⌈η̄⌉, βορρᾶ βήματ[α] ⌈ϛ̄⌉, ἀπηλιώτο[υ]
2   [*c.* 4–5] . ϛιδια π⟨ή⟩χεις ῑε λιβὸς ῑγ. εὑρεῖν τὰς ἀρούρα[ς.]
3   [οὕ]τω ποιοῦμεν. συντίθω τὸ ν̣ότον καὶ τὸν βορέα,
4   [η̄] κ̣αὶ ϛ̄. γί(νεται) ῑδ. ὧν ἥμισυ ζ̄. ἐπὶ τὸν β̄. διὰ τί ἐπὶ τὸν β̄; [ὅτι]
5   τὸ βῆμα ἔχει πήχεις β̄. γί(νεται) ῑδ πήχεις. καὶ συντίθω
6   [τ]ὸ̣ν ἀπηλιώτον καὶ τὸν λιβάν, ῑε καὶ ῑγ. γί(νεται) κ̄η. ὧν
7   [ἥ]μισυ ῑδ. ἐπὶ τοῦ πήχεις. γί(νεται) π̄δ. π̄δ ἐπὶ τὸν ῑδ.
8   γί(νεται) ['Ά]ρ̣[ο]ϛ̣. λοιπαὶ Άροϛ̣. ⟨παρὰ⟩ τὸ Ὄϛιϛ̣. γί(νεται) (symbol) τ̄π̄δ̄.
9   [ἔσ]ται ἄρουραι (symbol) τ̄π̄δ̄. οὕτως ἔχει‶ ὁμοίως.
10   *decorative border*

11   ὀρθογώνιον, ὑποτίνουσα ῑζ. εὑρεῖν τὰς ἄλλας
12   [π]λευράς. οὕτω ποιοῦμεν. τὰ ῑζ τῆς ὑποτινούσα′
13   [ἐ]φ' ἑαυτά̣. γί(νεται) σ̄π̄θ. εἰς δύο πλευρὰς ἔσται. η̄ ἐφ' ἑαυτά.
14   γί(νεται) ξ̄δ. ἀπὸ τῶν σ̄π̄θ οἰφέλομεν ξ̄δ. λοι⟨πὰ⟩ σκε. ὧν
15   [πλ]ευρὰ ῑε. ἄρα ἦν ὀρθὴ η̄, βάσις ῑε, ὑποτίνουσα
16   [ῑζ.] οὕτως ἔχει ὁμοίως.
17   *paragraphos, diagram* μ | ὀρθὴ η | βάσις | ιε | ὑποτίνουσα ῑζ

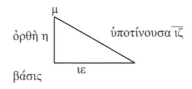

ὀρθὴ η   ὑποτίνουσα ῑζ

βάσις   ιε

18   [χ]ωρίων ἀμπέλιον, νότος ἐπὶ βορρᾶ πήχ-
19   [εις] λ̄, ἀπὸ δὲ λιβὸς εἰς ἀπηλιώτου πήχεις κ̄.
20   εὑρεῖν πόσους ἀμπέλους χωρήσε. λ̄ ⟨ἐ⟩πὶ τὸν κ̄.
21   [γί(νεται)] χ″. τοῦτον τὸ τέταρτον. γί(νεται) ρ̄ν′. ἄρα ὁ χωρίων
22   ἀμπέλων ρ̄ν. οὕτως ἔχει ὁμοίως.⁗
23   *paragraphos, decorative border*

2 πήχεις: *error for* κάλαμοι | 3 *l.* νότον | *l.* βορρᾶν? | 6 *l.* ἀπηλιώτην | 7 τοῦ: *l.* τοὺς | 8 (symbol) τ̄π̄δ̄: *l.* η′ τπδ′ | 9 (symbol) τ̄π̄δ̄: *l.* η′ τπδ′ | 11 *l.* ὑποτείνουσα | 12 *l.* ὑποτεινούσης | 14 *l.* ὑφέλομεν | 15 *l.* ὑποτείνουσα | 17 *l.* ὑποτείνουσα | 18 *l.* χωρίον | 19 *l.* ἀπηλιώτην | 20 *l.* χωρήσει | 21 *l.* τούτων | *l.* χωρίον | 22 *l.* ἀμπέλους

F verso

(f5: mathematical problem)

- - - - -

1   ... [on the south] 8 paces, on the north 6 paces, on the east

2   ... 15 cubits (*error for* reeds), on the west 13. To find the arourai.

3   We proceed as follows. I add the south and the north (sides),

4   [8] and 6. The result is 14. Half of this is 7. Times 2. Why times 2? [Because]

5   one pace contains 2 cubits. The result is 14 cubits. And we add

6   the east and the west, 15 and 13. The result is 28. Half of this

7   is 14. (I convert?) into cubits. The result is 84. 84 times 14.

8   The result is 1[1]76. The remainder (*error for* result) is 1176. \<Divided by\> 9216. The
result is (symbol) 1/384 (*error for* ⅛ 1/384).

9   It will be (symbol) 1/384 arourai. This way for similar cases.

10   *decorative border*

(f6: mathematical problem)

11   A right-angled (triangle) whose hypotenuse is 17. To find the other

12   sides. We proceed as follows. The 17 of the hypotenuse

13   times itself. The result is 289. It will be into two sides. 8 times itself.

14   The result is 64. We subtract 64 from 289. The remainder is 225. The square root

15   of this is 15. Hence the vertical was 8, the base 15, the hypotenuse

16   [17.] This way for similar cases.

17   *diagram*

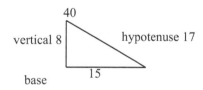

(f7: mathematical problem)

18   A field for vines, from south to north 30 cubits,

19   from west to east 20 cubits.

20   To find how many vines it will hold. 30 times 20.

21   [The result is 600.] One quarter of this. The result is 150. Hence the field

22   (will hold) 150 vines. This way for similar cases.

23   *decorative border*

G recto

1   μέτρον εἴ[δη] ἐστὶν τάδε·ʹ δάκτυλ[ος,ʹ παλαιστής,ʹ]
2   λιχάς,ʹ ψιθαμή,ʹ πούς,ʹ πηγών,ʹ πῆχ[υς,ʹ βῆμα,ʹ]
3   ξύλον,ʹ ὄργηεια,ʹ κάλαμος,ʹ ἄκενα,ʹ ἄμμ[α,ʹ πλέθρον,ʹ]
4   σχοινίον,ʹ στάδιον,ʹ δίαυλον,ʹ δόλιχος,ʹ γ[   ]
5   μείλιον,ʹ παρασάγης,ʹ σχοῖνος,ʹ ⟨ἢ⟩ δωδεκατ[ικὸν]
6   καλῖται. ἐστὶν λα῾ο῾ύρα,ʹ ἄμφοδος,ʹ ὄργοιεια̣ʹ [c. 5–7 τῶ-]
7   νδε τῶν μέτρον δάκτυλοςʹ ὁ προάγωνʹ κ̣[αὶ ἐλά-]
8   χιστος,ʹ καθάπερ καὶ ἡ μωνάς,ʹ ἥτις οὐκ [   ]
9   ἄλλως.ʹ σύνκιτε ἡ τῶν ἰδίων μέτρω̣ν̣ [μέ-]
10   τρων παληστής.ʹ οἱ β̄ παλισταὶ λιχάς·ʹ οἱ [γ̄ παλαισ-]
11   ταὶ ψιθαμή·ʹ ὁ δ̄ παλησταὶ πούς πτολεμαικ[ός·]
12   οὗτός ἐστιν ὁ Ἐκύπτιος.ʹ ὁ γὰρ Ἰδαλικὸς πού̣[ς]
13   δακτύλων ἐστὶν ι̅γ̅ γʹʹ· ὁ δὲ ⟨τεκτο⟩νεικὸς πούς δ̣[ακτύλω-]
14   ν ἐστὶν ι̅γ̅ ω̅ʹ· διεροῦντε δὲ καὶ αὐτοὶ ἰς δισι[   ]
15   της δ. οἱ πέντε παλεσταὶςʹ πῆχειςʹ λιγ[οΰφικός·]
16   καλῖτη δὲʹ καὶ ποιγών·ʹ οἱ ϛ̄ παληισταί πῆχει[ς τεκτονι-]
17   κός·ʹ ὁ δὲ αὐτὸςʹ καὶ δημόσιος καλ̣[εῖται·]
18   οἱ ζ̄ π̣α̣ληισταὶ πῆχεις νιλομετρικός· [οἱ η̄ παλαισ-]
19   ται πῆχεις εἰστονικός·ʹ ο῾ί̅ʹ θ̄ παληισ[ταὶ   ]
20   τος ἐστὶν ἡ διάστασις τῶν σκελῶν.ʹ [   ]
21   *decorative border*

22   ναύβιον στρονκύλουν μισ.ων[περιφέρεια]
23   πηχῶν κ̄, βάθος πηχῶν α̅ S̅ʹ d̅ʹ. εὑρ[εῖν πόσα   ]
24   χωρήσι. οὕτω ποιοῦμεν. λαμβάνω τὸ ἥμ[ισυ τῆς περιφερ-]
25   ίας. γί(νεται) ι. ι̅[[ε]] ἐφ᾽ ἑαυτά. γί(νεται) ρ̅. τούτων τ[ὸ ιβʹ. γί(νεται)]
26   η̅ γʹʹ. ἐπὶ τὸ βά⟨θ⟩ος, πηχῶν α̅ S̅ʹ d̅ʹ. γί(νεται) [ι̅δ S̅ʹ ιβʹ. οὕ-]
27   τ̣ως ἔχει ὁμοίως.   *line filler*
28   *diagram* ιδ επ ι̅. | α̣ Sʹ ḍ | *traces*

G recto

(g1: metrological text)

1   The kinds of measures are as follows: finger, [palm,]
2   lichas, spithame, foot, pygon, cubit, [pace,]
3   xylon, fathom, reed, akaina, hamma, [plethron,]
4   schoinion, stade, diaulon, dolichos, . . .
5   mile, parasang, schoinos, (which) is called dodekatikon.
6   They are laura, amphodon, fathom, . . .
7   Of these measures, the leading one and least one is the finger,
8   just like the unit, which . . . not . . .
9   otherwise. The . . . of the particular measures comprises . . .
10   of the measures the palm. 2 palms are a lichas. [3] palms
11   are a spithame. 4 palms are a Ptolemaic foot.
12   This is the Egyptian (foot). For the Italian foot
13   comprises 13⅓ fingers. The builder's foot comprises
14   13⅔ fingers. They are also divided into . . .
15   4. 5 palms are a weaver's cubit.
16   It is also called a pygon. 6 palms are a builder's cubit.
17   The same is also called a public (cubit).
18   7 palms are a nilometric cubit. [8 palms are]
19   a loom cubit. 9 palms are . . .
20   is the distance between the legs. . . .
21   *decorative border*

(g2: mathematical problem)

22   A circular trench . . . [circumference]
23   20 cubits, depth 1½ ¼ cubits. To find [how many . . .]
24   it will hold. We proceed as follows. I take half of the circumference.
25   The result is 10. 10 times itself. The result is 100. [1/12] of this. [The result is]
26   8⅓. Times the depth, 1½ ¼ cubits. The result is [14½ 1/12.] This way
27   for similar cases.

G recto (continued)

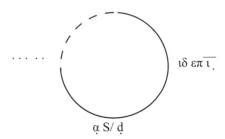

1 *l.* μέτρων | 2 *l.* σπιθαμή | *l.* πυγών | 3 ὄργηεια: ορθη⸍εια⸍ *sed* θ *corr. in* γ, *l.* ὄργυια | *l.* ἄκαινα | 5 *l.* μίλιον | *l.* παρασάγγης | 6 *l.* καλεῖται | *l.* λαύρα | *l.* ἄμφοδον | *l.* ὄργυια | 7 *l.* μέτρων | 8 *l.* μονάς | 9 *l.* σύγκειται | ϊδιων *pap.* | 10 *l.* παλαιστής | 11 *l.* σπιθαμή | *l.* οἱ | *l.* παλαισταὶ | 12 *l.* Αἰγύπτιος | ϊδαλικος *pap.*, *l.* Ἰταλικὸς | 13 *l.* τεκτονικὸς | 14 *l.* διαιροῦνται | ϊς *pap.*, *l.* εἰς | 15 *l.* παλαισταὶ | *l.* πῆχυς | 16 *l.* καλεῖται | *l.* πυγών | *l.* παλαισταί | *l.* πῆχυς | 18 *l.* παλαισταὶ | *l.* πῆχυς | 19 *l.* πῆχυς ἱστονικός | *l.* παλαισταὶ | 22 *l.* στρογγύλον | 24 *l.* χωρήσει | 24–25 *l.* περιφερείας | 25 ι¹: ϊ *pap.*

G recto (continued)

28   *diagram*

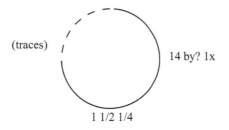

G verso

- - - - -

1   [. . . . .] . . . . μ . . ο . [.] . ἰ[σο]σκελὴς ειοσυ . [   ]

2   [ἰσοσκελὴς] ἀνὰ σχοινία κ̄, κυνὴ [β]άσις λ̄β̄. εὑρεῖν τὰς ἄλλα[ς]

3   [πλευρ]άς. οὕτω ποιοῦμεν· λαμβάνομαι τὸ ἥμισυ τῆς

4   [κοινῆ]ς βάσεως, ῑς̄. ἐφ᾽ ἑαυτά. γί(νεται) σ̄ν̄ς̄. καὶ τὰ κ̄ ἐφ᾽ ἑαυτά. γί(νεται) ῡ.

5   [ἀπὸ τῶ]ν υ ἠφέλομεν σ̄ν̄ς̄. λοπαὶ ρ̄μ̄δ̄. ὧν πλευρὰ ῑβ.

6   [ἄρα ἦν] ἡ ἑκάστη ὀρθὴ ῑβ. εὑρεῖν τὸ ἐνβαδόν. τὴν βάσιν

7   [ἐπὶ τὴν] ὀρθήν, ῑβ ἐπὶ τὸν ῑς̄. γί(νεται) ρ̄ϙβ. ὧν ἥμισυ ϙ̄ς̄.

8   [ἄρα ἦν τ]ὸ ἐνβαδὸν ἀρουρῶν ϙ̄ς̄. οὕτως ἔχει ὁμοίως.⫽

9   *diagram* [ἄρα ἦ]ν ἑκάστη ὀρ(θὴ)′ | ῑβ | ρ̄ϙβ | ῑς̄ | ϙ̄ς̄ | κοινοὶ βάσις λ̄β̄ | ῑς̄

10   [. . . . .]ν στρονκύλουν ἴσον, μῆκος πηχῶν λ̄,

11   [πλάτος δ]ακτύλων η̄, πάχος δακτύλων δ̄. εὑρεῖν

12   [τὸ στ]ερεόν. οὕτω ποιοῦμεν. τὸ πλάτος ἐπὶ τὸ

13   [μῆκος, τὸ] η̄ ἐπὶ τὸ λ̄. γί(νεται) σ̄μ. ταῦτα ἐπὶ τὸ πάχος,

14   [τὸ σ̄μ ἐπὶ τὸ δ̄.] γί(νεται) λ̄ξ. ταῦτα μερίζομαι παρὰ τὸν

15   [σπη, καὶ τὰ λοιπ]ὰ εἰς δακτύλους {ῑβ}. γί(νεται) γ̄ καὶ δακ-

16   [τύλους η̄. οὕτ]ως ἔχει ὁμοίως. *line filler*

17   *diagram* . | πῆχος γ̄′ δα̣κ̣ | η̄

18   *decorative border*

2 *l.* κοινὴ | 5 ϋ *pap.*, *l.* ὑφέλομεν | *l.* λοιπαὶ | 6 *l.* ἐμβαδόν | 8 *l.* ἐμβαδὸν | 9 *l.* κοινὴ | 10 *l.* στρογγύλον | 17 *l.* πήχεις?

G verso

(g3: mathematical problem)

- - - - -

1   … isosceles …
2   isosceles having (legs) 20 schoinia each, common base 32. To find the other
3   sides. We proceed as follows. I take half of the
4   common base, 16. Times itself. The result is 256. And 20 times itself. The result is 400.
5   We subtract 256 [from] 400. The remainder is 144. The square root of this is 12.
6   [Hence] each vertical [was] 12. To find the area. The base
7   [times the] vertical, 12 times 16. The result is 192. Half of this is 96.
8   [Hence] the area [was] 96 arourai. This way for similar cases.
9   *diagram*

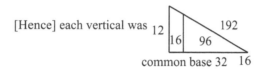

(g4: mathematical problem)
10   A circular equal? trench?, having length 30 cubits,
11   [breadth] 8 fingers, thickness 4 fingers. To find
12   [the] volume. We proceed as follows. The breadth by the
13   [length,] 8 times 30. The result is 240. This times the thickness,
14   [240 times 4.] The result is 960. I divide this by
15   [288, and (I convert) the remainder] into fingers. The result is 3 and [8] fingers.
16   This way for similar cases.
17   *diagram*

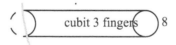

18   *decorative border*

M recto

- - - - -

1   [*c. 14*] σχοιν[ι]α . . . . [   ]
2   [*c. 14*]ον[.]ανεχονια . ρ[. .] . . . [   ]
3   [. . . .] ἡ βαρβαρικὴ σχῦνος θ . . εστινημουσου[   ]
4   [καλ]ουμένης σχοινία. τὰ σ̅μ̅ σχοινία σχ[οῖνος, ἣ]
5   [δω]δεκατικὸν καλοῦντε. ἐστὶν λαύρα ἡ μυρ[ιὰς]
6   πηχῶν. ἄμφοτον ἔχει τὸ μὲν μοῖκος π[ηχῶν σ̅,]
7   τὸ δὲ πλάτος πηχῶν ρ̅· γί(νεται) ἐνβαδοὶ πηχῶν [μυριάδες]
8   δύο″. τὸ καλούμεν⟨ον⟩ δωδεκατικόν ἐστιν ἀμφ[όδων ρ̅,]
9   λαοωρῶν δὲ ρ̅π̅. τὰ ναύβια ἐκ πανταχόθεν [ἐστὶν?]
10   ξύλων ἕν, βάθος ξύλων ἕν. τὸ δὲ ξύλων ἔχει πήχ[εις]
11   γ, ὥστε εἶναι τὸ μὲν τυμόσιον ναύβιον στερε[ὰς]
12   πήχης κ̅ζ̅. ὁ στερεὸς πῆχεις χωρῖ ξυροῦ ἀ(ρτάβας) γ̅ d̅ [η̅],
13   ἡκροῦ ταὶ μετρήτας ⟨γ⟩. ἡ ἄρουρα ἐστὶν· ἡ μὲν κατ' ἄκρ[ον]
14   ἔχει βίκους μ̅η̅, ἡ δὲ κατὰ πόλιν ν̅. ἔχει ⟨ὁ⟩ σχοῖν[ο]ς [μίλια]
15   δ̅. τὸ δὲ μίλιον ἔχει γοίας γ̅. τὸ τὲ γοίης ἔχ[ει στάδια]
16   ε̅. τὸ δὲ στάδιον ἔχει σχοινία γεωμετρικὰ δ̅, τ[ὸ δὲ σχο-]
17   ινίων τὸ γεωμετρικὸν ⟨ἔχει⟩ πωχῶν ο̅ϛ̅, ὥστε [ἔχειν]
18   τὴν σχοῖν⟨ον⟩ [γ]οίας μὲν ι̅β̅, στάδιον ξ̅, σχοινία γ[εω-]
19   μετρικὰ σ̅μ̅, πηχ{ηχ}ῶν {δ̅} (μυριάδας) β̅ ῾Γ̅μ̅ ξ[. .] . νδε . [
20   *decorative border, palm fronds, and ankh*

3 *l.* σχοῖνος | θ *corr. ex* γ? | 5 *l.* καλοῦνται, *melius* καλεῖται, *cf.* Gr. 6 | 6 *l.* ἄμφοδον | *l.* μῆκος | 7 *l.* ἐμβαδῶν | 9 *l.* λαυρῶν | *l.* τὸ ναύβιον | 10 ξύλων *l.* ξύλον (*ter*) | ἕν² : ε̅ν̅ *pap.* | ξύλων³ : ξ *corr. ex* φ? | ἔχει: χ *corr. ex* κ | 11 *l.* δημόσιον | 12 πήχης: *l.* πήχεις | πῆχεις: *l.* πῆχυς | *l.* χωρεῖ | *l.* ξηροῦ | 13 *l.* ὑγροῦ δὲ | *l.* ἄγρον | 15 τὸ *l.* ὁ | τὲ: *l.* δὲ | *l.* γύας | *l.* γύης | 16 γεωμετρικὰ: ικ *corr. ex* υ | 16–17 *l.* σχοινίον | 17 *l.* πηχῶν | 18 *l.* γύας | *l.* στάδια | .νδε. : ν *corr.*

M recto

(m1: metrological text)

- - - - -

1   . . . schoinia . . .
2   . . .
3   . . . the barbarian schoinos . . .
4   so-called . . . schoinia. 240 schoinia (are) called a schoinos [or]
5   dodekatikon. A laura comprises 10,000
6   cubits. An amphodon contains in length [200] cubits,
7   in breadth 100 cubits; it comprises 20,000 area cubits.
8   The so-called dodekatikon comprises [90] amphoda,
9   and 180 laurai. The naubion [is?] in all directions
10   one xylon, and one xylon in depth. The xylon contains 3 cubits,
11   so that the public naubion comprises
12   27 solid cubits. The solid cubit holds 3 ¼ [⅛] artabas of dry goods,
13   and ⟨3⟩ metretai of fluid. The aroura is: the country aroura
14   contains 48 bikoi, and the urban (aroura), 50 (bikoi). The schoinos contains
15   4 [miles.] The mile contains 3 guai. The gues contains 5 [stades.]
16   The stade contains 4 surveyor's schoinia, and the
17   surveyor's schoinion ⟨contains⟩ 96 cubits, so that
18   The schoinos contains 12 guai, 60 stades, 240 surveyor's schoinia,
19   23,040 cubits. . . .
20   *decorative border, palm fronds, and ankh*

M verso

1   [   ]ς ὑπερίχει [*c. 4*] ̣ ̣[*c. 3*] ̣ ̣[   ]
2   [*c. 7*]ε. οὕτω ποιοῦμεν. τοὺς γενωνοὺ[ς, ε̅,]
3   [ἐφ' ἑαυτ]ᾳ̣. γί(νεται) κ̅ε̅. ἄλλα καὶ πέντε. γί(νεται) λ̅. ὧν ἥ[μισυ]
4   [ι̅ε̅. ἐπὶ] τὸν η̅. γί(νεται) ρ̅κ̅. ἀπὸ τῶν τ ἠφέλω`σ´μεν ̅[ρ̅κ̅.]
5   [γί(νεται)] ρ̅π̅. παρὰ τὸν τοὺς κυνωνούς. γί(νεται) λ̅[ϛ̅.]
6   [προστίθ]ωμεν ἄλλας η̅. γί(νεται) [μ̅]δ̅. ἄρα ὁ πρῶτ[ο]ς
7   [λήμψ]ετε ποιροῦ ἀ(ρτάβας) μ̅δ̅. προστίθωμεν ἄλλας η̅.
8   [γί(νεται) ν̅β̅.] ἄρα ὁ τεύτερος λήμψετε ποιροῦ ἀ(ρτάβας) ν̅β̅. καὶ
9   [π]ροστίθωμεν η̅. γί(νεται) ξ̅. ἄρα ὁ τρίτ̣[ο]ς λήμψετε
10   ποιροῦ ἀ(ρτάβας) ξ̅. καὶ προστίθωμεν η̅. γί(νεται) ξ̅η̅. ἄρα
11   [ὁ] τέταρτος λοίμψετε ποιροῦ ἀ(ρτάβας) ξ̅η̅. καὶ προστίθω-
12   μ[εν] η̅. γί(νεται) ο̅ϛ̅. ἄρα ὁ πένπτος λήμψετ̣[ε] ποιροῦ
13   [ἀ(ρτάβας) ο̅]ϛ̅. συντίθω τὰς ἀρτάβας, μ̅δ̅ καὶ ν̅β̅ κ̣αὶ ξ̅ καὶ
14   [ξ̅η̅ καὶ ο̅]ϛ̅. γί(νεται) τ̅. οὕτως ἔχει ὁμοίως. //
15   *diagram* μδ | νβ | ξ | ξη | οϛ | τ η

μδ
_____

νβ   τ η
_____

ξ
_____

ξη
_____

οϛ
_____

16   *palm fronds and ankh*

1 *l.* ὑπερέχει | 2 *l.* κοινωνούς | 4 *l.* ὑφέλομεν | 5 *l.* τῶν κοινωνῶν | 6 *l.* προστίθομεν | 7 *l.* λήψεται πυροῦ | *l.* προστίθομεν | 8 *l.* δεύτερος λήψεται πυροῦ | 9 *l.* προστίθομεν | *l.* λήψεται | 10 *l.* πυροῦ | *l.* προστίθομεν | 11 *l.* λήψεται πυροῦ | 11–12 *l.* προστίθομεν | 12 *l.* πέμπτος λήψεται πυροῦ

M verso

(m2: mathematical problem)

1   . . . exceeds . . .
2   . . . We proceed as follows. The (number of) sharers, [5,]
3   [times] itself. The result is 25. Another 5. The result is 30. [Half] of this is
4   [15. Times] 8. The result is 120. We subtract [120] from 300.
5   [The result is] 180. (I divide) by the (number of) sharers. The result is 3[6.]
6   We add another 8. The result is [4]4. Hence the first (sharer)
7   will get 44 artabas of wheat. We add another 8.
8   [The result is 52.] Hence the second (sharer) will get 52 (artabas) of wheat. And
9   we add 8. The result is 60. Hence the third (sharer) will get
10   60 artabas of wheat. And we add 8. The result is 68. Hence
11   [the] fourth (sharer) will get 68 artabas of wheat. And we add
12   8. The result is 76. Hence the fifth (sharer) will get
13   [7]6 [artabas] of wheat. I add the artabas, 44 and 52, and 60 and
14   [68 and 7]6. The result is 300. This way for similar cases.
15   *diagram*

44	
52	300 8
60	
68	
76	

16   *ankh*

N recto

- - - - -

1   [κ]άτω δ[ιάμετρο]ς πηχ[ῶν λ̅β̅, βάθος]
2   [πηχῶ]ν π̅. εὑρεῖν πόσα ναύβια ἀ̣ναιβ[λήθη]
3   [οὕτ]ῳ ποιῶ. συντίθω τὰς δύο διαμέτρ[ους, τουτ-]
4   [έσ]τιν μ̅ καὶ λ̅β̅. γί(νεται) ο̅β̅. ὧν ἥμησυ γί(νεται) λ̅ς̅. [λ̅ς̅]
5   [ἐφ' ἑα]υτά. γί(νεται) ʹΑσϙς̅. τούτων ἐφέλω τ̣[ὸ] δ̅. λο[ιπαὶ λ̅ο̅β̅.]
6   [λ̅ο̅β̅] ἐπ[ὶ τ]ὸ βάθος, πηχῶν π̅. γί(νεται) [(μυριάδες)] ζ̅ ʹΖψξ̅.
7   [τα]ῦτα μερίσω παρὰ τῶν κ̅ζ̅. διότι; [ἔχει] τὸ ναύ-
8   [β]ιον πήχεις κ̅ζ̅. γί(νεται) Β̅ω̅π̅. ἄρα [ἀνεβ]λήθη
9   [ν]αύβια Β̅ω̅π̅.
10   *diagram* μ | λβ | π | ʹΒωπ

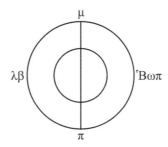

11   *decorative border*

12   θησαυρὸς τρίγωνος, μῆκος πηχῶν κ̅, πλάτος πηχῶν
13   λ̅, βάθος πηχῶν ς̅. εὑρεῖν τὰς ἀρτάβας. οὕτω ποιοῦμεν.
14   [τ]ὸ πλάτος ἐπὶ τὸ μῆκος, κ̅ ἐπὶ τὸν λ̅. γί(νεται) χ̅. τούτων
15   τὸ τέταρτον. γί(νεται) ρ̅ν̅. ἐπεὶ δὸ βάθος, πηχῶν ς̅.
16   [γί(νεται)] λ̅. λ̅ ἐπὶ τὸν γ̅ d̅ η. γί(νεται) ʹΓλζ̅ S′. οὕτως ἔχει.″
17   *diagram* κ | λ | ς | ʹΓλζ̅ S′

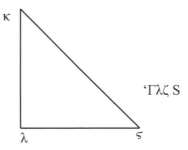

18   *palm frond and decorative border*

2 *l.* ἀνεβλήθη | 4 *l.* ἥμισυ | 5 *l.* ὑφέλω | 7 *l.* μερίζω? | τῶν *l.* τὸν | 15 *l.* ἐπὶ τὸ | 16 | ʹΓλζ̅ S′ *pap.*,
ʹΓ *corr. ex* S′?

N recto

(n1: mathematical problem)

     - - - - -

1   . . . lower diameter [32] cubits, [depth]

2   80 cubits. To find how many naubia were excavated.

3   I proceed as follows. I add the two diameters, that is

4   40 and 32. The result is 72. Half of this is 36. [36]

5   [times] itself. The result is 1296. I subtract ¼ from this. The remainder is [972.]

6   [972] times the depth, 80 cubits. The result is 77,760.

7   I divide this by 27. Why? (Because) one naubion [contains]

8   27 cubits. The result is 2880. Hence

9   2880 naubia were excavated.

10   *diagram*

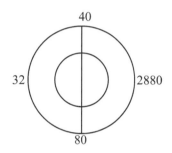

11   *decorative border*

(n2: mathematical problem)

12   A triangular granary, having length 20 cubits, breadth 30 cubits,

13   depth 6 cubits. To find the artabas. We proceed as follows.

14   The breadth by the length, 20 by 30. The result is 600.

15   One quarter of this. The result is 150. Times the depth, 6 cubits.

16   [The result is] 900. 900 times 3 ¼ ⅛. The result is 3037 ½. This way (for similar cases.)

17   *diagram*

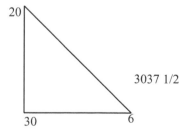

18   *palm frond and decorative border*

N verso

- - - - -

1   [εὑρε]ῖ[ν] πόσ[ας ἀ]ρτ[ά]βας χώρη[σει ὁ θησαυ-]
2   ρ[ός. οὕ]το π[ο]ιοῦμεν. [πο]λυπλαδιάζω[μεν τὸ μῆκος]
3   ἐπ[ὶ] τὸ πλάτος, κ̄ [ἐ]πὶ τὸν ῑ. γί(νεται) σ̄. ὧ[ν ἥμισυ ρ̄.]
4   ταῦτα ποιῶ ἐπὶ τὸ βάθος, πηχῶ[ν γ̄. γ(ίνεται) τ̄. τ̄]
5   ἐπὶ τὸν γ̄ d̄ η̄. γί(νεται) Ἁιβ S̷. ἄρα χωρήσι ὁ [θησαυρὸς]
6   σίτου ἀρτάβας Ἁιβ S̷. καὶ ἐπὶ τῶν ὁμοί[ων.]
7   *diagram* ἀ(ρτάβαι) Ἁιβ S̷

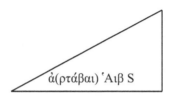

ἀ(ρτάβαι) Ἁιβ S

8   θησαυρὸς καμαρωτὸς σχοῖμα . . . . .[   ]
9   μενος ὑπὸ τεσάρων διαστάσεω[ν] . . .[   ]
10   τὸ μῆκος πηχῶν κ̄ε̄, τὸ τὲ πλά[τ]ος π[ηχῶν]
11   ῑε̄, τὴ δὲ καμάρα πηχῶν ῑη̄, βάθος [πηχ-]
12   ῶν ῑϛ̄. εὑρεῖν πόσας ἀρτάβας χωρήσι ὁ θ[ησ-]
13   αυρός. οὕτω ποιῶ. ποληπλαδιάζω τὸ [μῆκος]
14   ἐπὶ τὸ πλάτος, τούτεστιν κ̄ε̄ ἐπὶ τὸν ῑε̄. γί(νεται) [τοε.]
15   ὁμοίως πολυπλαδιάζω τὴν καμάραν ἐ[πὶ]
16   τὸ βάθος, ῑϛ̄ ἐπὶ τὸν ῑη̄. γί(νεται) σ̄π̄η̄. ταῦτα [πολυ-]
17   πλαδιάζω ἐπὶ τὸν τοε. γί(νεται) (μυριάδες) ῑ Ἡ̄. [ἐ]πὶ τ[ὸ γ̄ d̄ η̄.]
18   [γί(νεται)] (μυριάδες) λ̄ϛ̄ Δ̄φ̄. ἄρα χωρήσι {ἄρα χωρήσι } [ὁ] θησ[αυρὸς]
19   [ἀρ]τάβας (μυριάδας) λ̄ϛ̄ Δ̄φ̄.
20   *diagram* (μυριάδες) λ̄ϛ̄ Δ̄φ

N verso

(n3: mathematical problem)

- - - - -

1   To find [how many] artabas [the granary] will hold.
2   We proceed as follows. We multiply [the length]
3   times the breadth, 20 times 10. The result is 200. [Half] of this is [100.]
4   I multiply this by the depth, [3] cubits. [The result is 300. 300]
5   times 3 ¼ ⅛. The result is 1012 ½. Hence the [granary] will hold
6   1012 ½ artabas of grain. And (so) for similar cases.
7   *diagram*

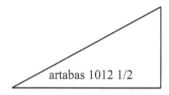

artabas 1012 1/2

(n4: mathematical problem)

8   A vaulted granary having shape ...
9   consisting? of four dimensions ...
10   the length 25 cubits, the breadth 15 cubits,
11   the vault 18 cubits, depth 16 cubits.
12   To find how many artabas the granary will hold.
13   I proceed as follows. I multiply the [length]
14   by the breadth, that is 25 by 15. The result is [375.]
15   Likewise I multiply the vault by
16   the depth, 16 by 18. The result is 288. I multiply this
17   by 375. The result is 108,000. Times [3 ¼ ⅛.]
18   [The result is] 364,500. Hence [the] granary will hold
19   364,500 artabas.

N verso (continued)

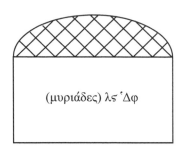

(μυριάδες) λϛ ʹΔφ

2 *l.* οὕτω | *l.* πολυπλασιάζομεν | 5 *l.* χωρήσει | 8 *l.* σχῆμα | 9 *l.* τεσσάρων | 10 *l.* δὲ | 11 τὴ: *l.* ἡ |
12 *l.* χωρήσει | 13 *l.* πολυπλασιάζω | 14 τούτεστιν: τ *corr. ex* δ | 15 *l.* πολυπλασιάζω |
16–17 *l.* πολυπλασιάζω | 18 *l.* χωρήσει | 19 *l.* ἀρταβῶν

N verso (continued)

20   *diagram*

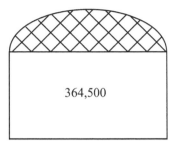

O recto

1   [θησαυρὸ]ς κ[α]μ[α]ρ[ω]τός, μῆκος π[ηχῶν ε̅, πλά-]

2   [τος π]ηχῶν γ̅, βάθος πηχῶν [β̅,] κ̣αὶ κ[α-]

3   μάρος πηχῶν β̅. εὐρὲν τὰς ἀρτάβας. οὕτο [ποιοῦμεν.]

4   τὸ̣ πλάτος ἐπ̣ὶ τὸ βά⟨θο⟩ς, ε̅ ἐπὶ τὸν β̅. γί(νεται) ι. ἐπ[ὶ τὸ πλάτος, γ̅.]

5   γί(νεται) λ̅. ἐπὶ τὴν καμάραν, β̅. γί(νεται) ξ̅. ἐπὶ τὸν γ̅ δ̅ η. [γί(νεται) σβ̅ S′.]

6   ἄρα χωρήσι ὁ θυσαυρὸς ἀρτάβας σβ̅ S′. ọ[ὕτως ἔχει ὁμοίως.]

7   *diagram* ε | γ | S′

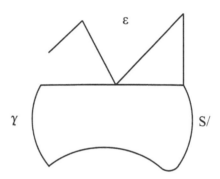

8   σφραγὶς, νότου [σχοι]νί[α] ὁσαδήποτε, βορρᾷ [δ̅,]

9   ἀπηλιώτου δ̅, λι[*vac.?*]βὸ̣ς δ̅, ἐπὶ ἀρούρας λβ̅. ọ[ὕτω]

10   ποι⟨οῦ⟩μεν. διπλοῦμεν τὰς ἀρούρας, διπλοῦμεν τὰ [λ̅β̅. γί(νεται) ξ̅δ̅.]

11   παρὰ τὸν δ̅. ⟨γί(νεται)⟩ ι̅ς̅. ἀπὸ τῶν ι̅ς̅ οἰφέλομεν δ̅. [γί(νεται) ι̅β̅.]

12   ἔσται νότου ι̅β̅. οὕτως ἔχει.////

13   *diagram* ιβ′ | δ | δ | δ | λβ

ιβ′   δ   δ
—————————————————————— λβ
δ

14   ψιλός, μῆκος πηχῶν ξ̅, πλάτος πηχῶ̣ν [λ̅ καὶ?]

15   πηχῶν ι̅β̅. εὑρεῖν πόσους βίκους ἔστ[αι.]

16   οὕτο ποιοῦμεν. λ̅ ἐπὶ τὸν ξ̅ γί(νεται) Ἀω̅. ἐπὶ

17   πηχῶν ι̅β̅. γί(νεται) (μυριάδες) β̅ Ἀχ̅. ταῦτα μερίζομ[εν]

18   παρὰ τὸν σ̅. διὰ τί παρὰ τὸν σ̅; ὅτι ὁ βῖκος

19   ἔχει ἐνβαδοὺς πήχεις σ̅′. γί(νεται) ρ̅η̅. ἄρα ἦν ọ̣

20   ψιλὼ βίκων ρ̅η̅. οὕτως ἔχει.//

21   *decorative border*

O recto

(o1: mathematical problem)

1   A vaulted [granary] having length [5] cubits,
2   [breadth] 3 cubits, depth [2] cubits, and vault
3   2 cubits. To find the artabas. [We proceed] as follows.
4   The breadth times the depth, 5 times 2. The result is 10. Times [the breadth, 3.]
5   The result is 30. Times the vault, 2. The result is 60. Times 3 ¼ ⅛. [The result is 202 ½.]
6   Hence the granary will hold 202 ½ artabas. This way [for similar cases.]
7   *diagram*

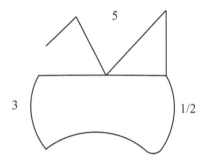

(o2: mathematical problem)

8   A plot, on the south some number of schoinia, on the north [4,]
9   on the east 4, on the west 4, (area) 32 arourai. We proceed as follows.
10   We double the arouras, we double [32. The result is 64.]
11   (I divide) by 4. ⟨The result is⟩ 16. We subtract 4 from 16. [The result is 12.]
12   The (side) on the south will be 12. This way (for similar cases.)
13   *diagram*

12	4	4
	4	32

(o3: mathematical problem)

14   A vacant lot having length 60 cubits, breadth [30] cubits [and?]
15   12 cubits. To find how many bikoi it will be.
16   We proceed as follows. 30 times 60. The result is 1800. Times
17   12 cubits. The result is 21,600. We divide this
18   by 200. Why by 200? Because one bikos
19   contains 200 area cubits. The result is 108. Hence the
20   building lot was 108 bikoi. This way (for similar cases).
21   *decorative border*

O recto (continued)

22   τρίκωνι καὶ πλευρὰ τὸν ι. οὕτω ποιοῦμε[ν. ὁ-]

23   κταπλοῦμεν τὰ δέκα. γί(νεται) π̄. προσθὲς ⌜π̄⌝ [ἐπὶ]

24   τὰν μίαν. γί(νεται) π̄ᾱ. ὧν πλευρὰ θ̄. οἰφέλωμ[εν]

25   τὰν μίαν′. λο`ι′παὶ η̄. τούτων τὸ τρίτον, β̄ ω̄. [οὕτως]

26   ἔχει ὁμοίως. *line filler*

2–3 *l.* καμάρα | 3 *l.* εὑρεῖν | *l.* οὕτω | 4 ï *pap.* | 6 *l.* χωρήσει | *l.* θησαυρὸς | 11 *l.* ὑφέλομεν | 16 *l.* οὕτω | 17 Ἀχ̄: χ *ex* ω | 19 ἐνβαδοὺς: α *inserted, l.* ἐμβαδοὺς | 20 *l.* ψιλὸς | 22 *l.* τρίγωνον | ï *pap.* | 24 *l.* τὴν | *l.* ὑφέλομεν | 25 *l.* τὴν

O recto (continued)

(o4: mathematical problem)

22   A triangle, and side 10. We proceed as follows.
23   We multiply by eight the ten. The result is 80. Add 80 [to]
24   one. The result is 81. The square root of this is 9. We subtract
25   the one. The remainder is 8. One third of this is 2⅔. [This way]
26   for similar cases.

O verso

- - - - -

1   [*c. 7* νότου σχ]οιν⟨ί⟩α ⌈η S′ d η⌉, βορρᾶ σχοιν[ία ϛ S′ d,]

2   [ἀπηλιώτ]ου σχοινία ε̅ S′ d η̅, λιβὸς σχοινία δ̣ S′.

3   [εὑρεῖ]ν τὰς ἀρούρας. οὕτου ποιοῦμεν. συντί-

4   [θωμεν τ]ὸ νῶτον καὶ τὸν βορέαν, η S̅′ d̅ η̅ καὶ ϛ S̅′ d̅.

5   [γί(νεται) ιε S̅′] {d̅} η. ὧν ἥμισυ {υ} ζ S̅′ d̅ ιϛ. καὶ πάλιν συντίθω-

6   [μεν τὸν] ἀπηλιώτου καὶ τὸν λίβαν, ε̅ S̅′ d̅ η καὶ δ̅ S′. γί(νεται) ι̅ S′.

7   [ὧν] ἥμισυ ε̅ ιϛ. τὰ ἥμισυ ἐπ' ἄλληλα. ἀπόδιξεις.

8   [ζ S̅′ d̅ ι]ϛ ἐπὶ τὸν η̅. γί(νεται) ξ̅β̅ S′. καὶ πάλιν ε̅ ι̅ϛ

9   [ἐπὶ τ]ὸν η̅. γί(νεται) μ̅β̅ S′. ξ̅β̅ S′ ἐπὶ τὸν μ̅β̅ S′. γί(νεται) Ἀφλα d.

10 [παρὰ τὸ]ν ξ̅δ. γί(νεται) λ̅θ̅ S̅′ λβ ξδ σνϛ. οὕτως ἔχει. ///

11 *diagram* [η] S′ d η | ε S′ d η | ϛ S′ d | δ S′ | λθ S′ [[d]] λβ ξδ σνϛ

[η] S′ d η	ε S′ d η	ϛ S′ d

δ S′

λθ S′ [[d]] λβ ξδ σνϛ

12 [. . . . .]ϛ ξυλοτομοῦντος ἐφ' ἡμέρας ι̅ε̅

13 [καὶ κο]μ̣ίσας ὑπὲρ μισθοὺς ἀργυρίου (τάλαντα) ρκ,

14 [καί εἰ]σὶ ἐρ[γά]της η̅ ἐργασζόμενη ἠφ' ἐμέρας ι̅ε̅.

15 [εἰ δὲ ϛ̅ εἰ]σί, π̣όσα λήμψονται ο ἵ ποταμῖται ὑπὲρ μ[ι]σ-

16 [θοῦ; ϛ̅] ἐπὶ τὸν ι̅ε. γί(νεται) ρ. καὶ η̅ ἐπὶ τὸν ι̅ε. γί(νεται) ρ̅κ.

17 [. . . .]μενοι ἀνὰ λόγον τὰ ρ̣, πολῖτε (ταλάντων) ρκ τὰ ρκ.

18 [τὰ ρ̅κ̅] ἐ[π]ὶ τον ρ̅κ. γί(νεται) (μυριὰς) α̅ Δ̅υ′. παρὰ τὸν η̅.

19 [γί(νεται) ρ̅]ξ. ἄρα λήμ]ψονται (τάλαντα) ρξ. οὕτως ἔχει ὁμοίως. //

20 *decorative border*

---

3 *l.* οὕτω | 4 *l.* τὸν νότον | *l.* τὸν βορρᾶν | 6 *l.* ἀπηλιώτην | *l.* λίβα | ῑ S̅′: *l.* ῑ η *error for* ῑ d̅ η | 7 *l.* ἡμίση | *l.* ἀπόδειξις | 9 μ̅β̅ S′ *(bis): error for* μ̅ S′ [8 × 5.0625 = 40.5] *uel rectius* μ̅α̅ S′ [8 × (10.375/2) = 41.5] | Ἀφλα d: *i.e.* 62.5 × 40.5! | 12 *l.* ξυλοτομοῦντες | 13 *l.* μισθοῦ | ρκ: κ *corr. ex* λ | 14 *l.* ἐργάται | ἐργασζόμενη: ζ *corr. ex* ομ (μ *incomplete*), *l.* ἐργαζόμενοι | *l.* ἐφ' ἡμέρας | 15 λήμψονται: αι *ex* ι?, *l.* λήψονται | ποταμῖται: π *remade* | μ[ι]σ-: μ *corr. ex* ε? | 16 ι̅ε: ε *corr. ex* ϛ | 17 ρ πολῖτε, *l.* ρ̣ πωλεῖται | 18 η̅ *error for* ρ̣ | 19 *l.* λήψονται

O verso

(o5: mathematical problem)

- - - - -

1 [on the south] 8 ½ ¼ ⅛ schoinia, on the north [6 ½ ¼] schoinia,
2 on the east 5 ½ ¼ ⅛ schoinia, on the west 4 ½ schoinia.
3 To find the arouras. We proceed as follows. We add
4 the south and the north (sides), 8 ½ ¼ ⅛ and 6 ½ ¼.
5 [The result is 15 ½] ⅛. Half of this is 7 ½ ¼ 1/16. And again we add
6 [the] east and the west (sides), 5 ½ ¼ ⅛ and 4 ½. The result is 10 ½ (*error for* 10 ⅛, *itself an error for* 10 ¼ ⅛).
7 Half [of this] is 5 1/16. The halves times each other. Demonstration.
8 [7 ½ ¼] 1/16 times 8. The result is 62 ½. And again 5 1/16 (*sic*)
9 [times] 8 is 42 ½ (*error for* 40 ½). 62 ½ times 42 ½ (*error for* 40 ½). The result is 2531 ¼.
10 [(I divide) by] 64. The result is 39 ½ 1/32 1/64 1/256. This way (for similar cases).
11 *diagram*

[8] 1/2 1/4 1/8   5 1/2 1/4 1/8   6 1/2 1/4
——————————————————————————————————————
4 1/2

39 1/2 1/32 1/64 1/256

(o6: mathematical problem)
12 […] brush-cutting for 15 days
13 [and] receiving for pay 120 talents of money,
14 [and] there are⁷ 8 workers working for 15 days.
15 [If, however], there are 6, how much will the river-workers take as pay?
16 [6] times 15. The result is 90. And 8 times 15. The result is 120.
17 … the 90 in proportion, 120 are sold for 120 talents.
18 [120] times 120. The result is 14,400. (I divide) by 8 (*error for* 90).
19 [The result is 160.] They will get 160 talents. This way for similar cases.
20 *decorative border*

H →

- - - - -

[–1]   [τὸ σχοινίον τὸ γεωμετρικὸν]

[0]   [ἔχει ἄμματα η̅, καλάμους ι̅ϛ̅, ξύλα λ̅β̅, βήματα]

1   [μ̅η̅, πήχε]ις ⌈ϙ̅ϛ̅⌉, πόδας ⌈ρ̅μ̅δ̅⌉, ψιθαμὰς ⌈ρ̅ϙ̅[β̅⌉, λιχάδας]

2   [σ̅π̅η̅,] π[αλ]εϲτὰς φ̅ο̅ϛ̅, δακτύλους ⟨Β⟩τ̅δ̅. τὸ δὲ ἄμμ[α]

3   ἔχει κ[αλ]άμου⟨ς⟩ β̅, ξύλα̣ δ̅, βῆμα ϛ̅, πήχεις ι̅β̅, πό[δας ι̅η̅,]

4   ψιθα[μ]ὰ̣ς κ̅δ̅, λιχ[ὰ]ς λ̅ϛ̅, παλεϲτὰς ο̅β̅, δακτύλ[ους σ̅π̅η̅. ὁ]

5   δὲ κ[άλα]μος ἔχει ξύλα β̅, βήματα γ̅″, πήχεις ϛ̅, [πόδας θ̅,]

6   [c. 5]θὰς ι̅β̅, λιχὰς ι̅η̅, παλεϲτὰς λ̅ϛ̅, δακτύλ[ους ρ̅μ̅δ̅.]

7   [τὸ δὲ ξ]ύ̣λ̣α̣ ἔχει βήματα α̅ S″, πήχεις γ̅, πόδας ⌈δ̅ [S″,⌉]

8   [ψιθαμ]ὰ̣ς ϛ̅, λιχάδας θ̅, παλεϲτὰς ι̅η̅, δακ[τύλους]

9   [ο̅β̅. τὸ δὲ] βῆμα ἔχει πήχεις β̅, πόδας γ̅, [ψιθαμὰς δ̅,]

10   [λιχὰς] ϛ̅, παλεϲτὰς ι̅β̅, δακτύλους μ̅η̅. ὁ δὲ πῆχ[υς]

11   [ἔχει πόδ]ας α̅ S′, ψιθαμὰς β̅, λιχάτας γ̅, παλεϲτὰς ̣[ ]

12   [δακτύλο]υς κ̅δ̅. ὁ δὲ ποὺς ἔχει ψιθαμὴν α̅ γ′,

13   [λιχάδας β̅,] παλεϲτὰς δ̅, δακτύλους ι̅ϛ̅. ἡ δὲ ψιθαμὴ

14   [ἔχει λιχ]άδας α̅ S′, παλεϲτὰς γ̅, δακτύλους ι̅β̅. ἡ δὲ [λιχὰς]

15   [ἔχει παλ]εϲτὰς {ο}β̅, δακτύλο̣υς η̅. ὁ τὲ παληϲτὴς

16   [ἔχει δακτύλο]υς δ̅. τὸ ναύβιον ἔχει ἄμματ[α β̅ d̅,]

17   [καλάμο]υς δ̅ S′, ξύλα θ̅, βήματα ι̅γ̅ S′, πήχεις κ̅ζ̅,

18   [πόδας μ̅ S′], ψιθαμὰ[ς ν̅]δ̅, λιχάδας π̅α̅, παλεϲτὰ[ς ρ̅ξ̅β̅,]

19   [δακτύλο]υς χ̅μ̅η̅. οὕτως. *line filler*

20   *decorative border*

21   [λοιπαὶ] ̅[.]̅, τὸ [. .] ἐν ⬚ μόρια· μὴ πρό-

22   [βα ρ̅. . . . . .] . . *line filler*

1 *l.* σπιθαμὰς | 2 *l.* παλαιστὰς | δακτύλους: ς *corr. ex* γ? | 3 *l.* βήματα | 4 *l.* σπιθαμὰς | *l.* λιχάδας | *l.* παλαιστὰς | 6 [c. 5]θας: *error for* σπιθαμὰς | ῑβ̅ *pap.* | *l.* λιχάδας | *l.* παλαιστὰς | 7 *l.* ξύλον | 8 *l.* σπιθαμὰς | *l.* παλαιστὰς | ῑη̅ *pap.* | 9 *l.* σπιθαμὰς | 10 *l.* λιχάδας | *l.* παλαιστὰς | ὁ *corr.* | 11 *l.* σπιθαμὰς | *l.* λιχάδας | *l.* παλαιστὰς | 12 *l.* σπιθαμὴν | 13 *l.* παλαιστὰς | *l.* σπιθαμὴ | 14 *l.* παλαιστὰς | 15 [παλ]εϲτὰς: *l.* παλαιστὰς | *l.* δὲ | παληϲτὴς: η² *ex* α, *l.* παλαιστὴς | 18 *l.* σπιθαμὰς | *l.* παλαιστὰς

H →

(h1: metrological text)

- - - - -

[–1]	[. . . The surveyor's schoinion]
[0]	[contains 8 hammata, 16 reeds, 32 xyla, 48 paces,]
1	96 cubits, 144 feet, 192 spithamai, [288 lichades,]
2	576 palms, ⟨2⟩304 fingers. The hamma [contains]
3	2 reeds, 4 xyla, 6 paces, 12 cubits, [18] feet,
4	24 spithamai, 36 lichades, 72 palms, [288] fingers. [The]
5	reed contains 2 xyla, 3 paces, 6 cubits, [9 feet,]
6	12 spithamai?, 18 lichades, 36 palms, [144] fingers. [The]
7	xylon contains 1½ paces, 3 cubits, 4 [½] feet,
8	6 spithamai, 9 lichades, 18 palms, [72] fingers.
9	[The] pace contains 2 cubits, 3 feet, [4 spithamai,]
10	6 [lichades,] 12 palms, 48 fingers. The cubit
11	[contains] 1½ feet, 2 spithamai, 3 lichades, … palms,
12	24 fingers. The foot contains 1⅓ spithame,
13	[2 lichades,] 4 palms, 16 fingers. The spithame
14	[contains] 1½ lichades, 3 palms, 12 fingers. The [lichas]
15	[contains] 2 palms, 8 fingers. The palm
16	[contains] 4 fingers. The naubion contains [2 ¼] hammata,
17	4½ reeds, 9 xyla, 13½ paces, 27 cubits,
18	[40½ feet,] [5]4 spithamai, 81 lichades, [162] palms,
19	648 fingers. This way (for similar cases.)
20	*decorative border*

(h2: mathematical problem)

21	[Remainder *n*,] 1/[*nn*] in [*n*] unit-fractions; do not
22	[surpass 100.] . . . .

H ↓

1 [Φλ]αουίῳ Διογένει ὑπα . . . . [ἀπὸ τῆς λαμ(πρᾶς)]
2 [καὶ] λαμ(προτάτης) Ὀξυρυγχιτῶν πόλε[ως] . . . .[. . . . . .]
3 [ἀ]πὸ ἐποικίου Ἱαιρέων ⳤ πάγ[ο]υ τῶ αὐτ[ο]ῦ νομ[οῦ ἑκου-]
4 [σί]ως ἐπειδέχομαι μισθώσασθαι ἀπὸ . .[. . . . . μα‶]
5 [καὶ] ι‵‶ ἀπὸ τῶν ὑπαρχόντων σοὶ . . . . .[. . . . . .]
6 [. .]‵ ἐκ βορᾶ τῶ χωρίου ἐδαφ . . θρ. . [. . . . .]
7 [ἀρ]ούρας δέσαρες ὥστε σπεῖραι ἐν μὲν λινωκ[αλάμη ἄρου-]
8 [ραν] μ[ί]ᾳν καὶ ἐν κριθὴν ἄρουραν μίαν καὶ [ἐν χόρ-]
9 [τω ἄρο]υραν δύο, φόρου τῆς λινωκαλάμη[ς λίνῳ]
10 [καθα]ρὰ εὐαρέστῳ δεσμίδια πενταμνειέα δώ[δεκα]
11 [καὶ] τῆς ἐν κριθῶν ἀνὰ κριθῶν ἀρτάβας π[έντε]
12 [κα]ὶ τῆς ἐν χόρτῳ ἐφ' ὑμησίας τῶ περεὶ κινῷ [　]
13 [. . .] ἀκίνδυνος ὁ φόρος παντὸς κινδύνου [　]
14 [πλη]ρῶσαι καὶ ἐπάναγκες ἀποδώσω . . [　]
15 [ἀν]υπερθέτως. κυρία ὑ ἐπειδοχή.
16 *decorative border, spiral, and palm frond*

1 ὕπα *pap.* | 3 ϊαιρεων *pap.*, *l.* Ἱερέων | *l.* τοῦ | 4 *l.* ἐπιδέχομαι | 5 ϊ‶ *pap.* | ὑπαρχοντων *pap.* | 6 *l.* βορρᾶ | *l.* τοῦ | 7 *l.* τέσσαρας | *l.* λινοκαλάμη | 9 *l.* ἀρούρας | *l.* λινοκαλάμης | 10 *l.* καθαρῷ | *l.* πενταμναῖα | 11 *l.* κριθαῖς | 12 *l.* ἡμισείας | *l.* περὶ κοινῷ | 14 *l.* πληρῶσαι | 15 ΰ *pap.*, *l.* ἡ | *l.* ἐπιδοχή

H ↓

(h3: model contract)

1   To Flavius Diogenes hyp- - - ... [from the splendid]
2   [and] most splendid city of the Oxyrhynchites .... [So and so son of so and so]
3   from the hamlet of Hiereon of the 6th pagus of the same nome [
4   I voluntarily undertake to lease from ... [- - - the 41st]
5   [and] 10th (year) from your property ... [
6   [. .] fields on the north of the vineyard ... [
7   four arouras, to sow one aroura in flax
8   and one aroura in barley and two arouras [in]
9   [hay], for a rent, for the flax, in [flax,]
10   twelve clean, good-quality five-mina bundles
11   [and] for the (aroura) in barley, at the rate of five artabas of barley,
12   and for the land in hay, for half-shares in common? [
13   [. . .] the rent being free of all risk [   ]
14   to pay, and I shall of necessity pay [   ]
15   without delay. The undertaking is authoritative.
16   *decorative border, spiral, and palm frond*

I →

- - - - -

[0]   [ἔχει ὁ πῆχυς]

1   [πα]ληστὰς ⌐Ϛ̅¬, ὁ [δὲ παλαιστὴς ἔχει δακτύλους δ̅, ὥστε]

2   [εἶν]αι τὸν πῆχων δακτύλ[ω]ν̣ [κδ̅. ἔχει ὁ μετρη-]

3   [τὴς] χόεις ιβ̅, ὁ δὲ χόευς ἔχει κοτύ[λας ιβ̅, ὥστε εἶναι τὸν]

4   [μετ]ρητὴν κοτυλῶ {μὲν} ρμδ̅. [ἔχει τὸ μναεῖον τε-]

5   [τάρτας] ιϛ̅, ἡ δετάρτη ἔχει θερμ[οὺς `μ]ὲν΄ [ϛ̅, κεράτια δὲ ιβ̅,]

6   [ὁ δὲ] θερισμοὺς ἔχει κεράτια β̅, [ὥστε εἶναι τὸ μναεῖον]

7   [θερ]μοὺς μὲν ϙϛ̅, κεράτιον δὲ ρϙβ̅. [ἔχει ἡ λίτρα οὐγκίας]

8   ιβ̅, ἡ δὲ ὀγεία ἔχει ἡμιόνκια μ[ὲν β̅, γράμματα δὲ]

9   κ̅δ̅, τὸ δὲ ἡμιόγια ἔχει γράμματα ⌐ι[β̅¬, ὥστε εἶναι τὴν λίτραν γράμματα]

10   σπη̅. ἔχει τὸ δάλαντα μνᾶς {εξ} ξ̅, ⟨ἡ⟩ δὲ μ[νᾶ ἔχει στατῆρας μὲν]

11   κε̅, δραχμὰς δὲ ρ̅, ὁ δὲ στατῆρα ἔχει δ[ραχμὰς μὲν δ̅,]

12   τετροβόλους δὲ ζ̅, ὥστε ε[ἶ]α̣ι τάλ[αντον στατήρων]

13   [μ]ὲν Ἀφ, δραχμῶν δὲ [ἑξακισχιλίων,]

14   [ὀ]β̣ο̣λῶν δὲ τ̣ε̣τρα̣[κισμυρίων δισχιλίων.]

15   *decorative border*

16   ληνὸς μῆκος πηχῶν ι̅η̅, πλ[άτος πηχῶν ϛ̅, βάθος]

17   πηχῶν ζ̅. εὑρεῖν τοὺς ξέστας. πο[ιοῦμαι. τὸ]

18   π⟨λ⟩άτος ἐπὶ τὸ μῆκος, ϛ̅ ἐπὶ τὸν [ι̅η̅. γί(νεται) ρη̅. ἐπὶ τὸ βάθος,]

19   πηχῶν ζ̅. γί(νεται) ψνϛ̅. γ̅ ἐπὶ τὸν [ψνϛ̅. διὰ τί ἐπὶ τὸν γ;]

20   ὅτι οἰγροῦ δὲ μητρητὰς γ̅. καὶ ἐπὶ [τὸν ψνϛ̅. γί(νεται) Βσξη̅.]

21   [c. 3] γ̅ ἐπὶ τὸν ο̅β̅. γί(νεται) σιϛ. ψ[νϛ̅ ἐπὶ τὸν σιϛ.]

22   [γί(νεται) (μυριάδες)] ι̅ϛ̅ Ϛσϙϛ. ἄρα ἦν χωρ[εῖ ἡ ληνὸς]

23   [ξεστῶν (μυριάδας) ι̅ϛ̅] Ϛσϙ[ϛ.] οὕτως ἔχει ὁ[μοίως.]

24   *decorative border*

1 *l.* παλαιστὰς | 2 *l.* πῆχυν | 3 χόευς: *l.* χοῦς | 4 *l.* κοτυλῶν | 5 *l.* τετάρτη | 6 *l.* θερμὸς | 7 *l.* κεράτια | 8 *l.* οὐγκία | *l.* ἡμιούγκια | 9 *l.* ἡμιούγκιον | 10 *l.* τάλαντον | μνασε[[ξ]]ξ̅δ̅ε *pap.* | 11 *l.* στατὴρ | 12 *l.* τετρωβόλους | 20 *l.* ὑγροῦ

I →

(i1: metrological text)

- - - - -

[0]	[. . . The cubit contains]
1	6 palms, the [palm contains 4 fingers, so that]
2	the cubit comprises [24] fingers. [The metretes]
3	[contains] 12 choeis, the chous contains [12] kotylai, [so that the]
4	metretes comprises 144 kotylai. [The mnaeion contains]
5	16 [tetartai,] the tetarte contains [6] thermoi [and 12 carats,]
6	[the] thermos contains 2 carats, [so that the mnaeion]
7	comprises 96 thermoi and 192 carats. [The litra contains]
8	12 [unciae,] the uncia contains [2] semiunciae and 24 [grammata,]
9	and the semiuncia contains 1[2] grammata, [so that the litra comprises]
10	288 [grammata]. The talent contains 60 minas, the mina [contains] 25 [staters,]
11	and 100 drachmas, and the stater contains [4] drachmas and
12	7 tetrobols, so that the talent comprises 1500 [staters,]
13	and [six thousand] drachmas,
14	and forty-[two thousand] obols.
15	*decorative border*

(i2: mathematical problem)

16	A vat having length 18 cubits, breadth [6 cubits, depth]
17	7 cubits. To find the xestai. I proceed (as follows). [The]
18	breadth times the length, 6 times [18. The result is 108. Times the depth,]
19	7 cubits. The result is 756. 3 times [756. Why times 3?]
20	Because (a solid cubit contains) 3 metretai of fluid. And times [756. The result is 2268.]
21	. . . 3 times 72. The result is 216. 7[56 times 216.]
22	[The result is] 163,296. Hence [the vat] holds
23	[16]3,29[6 xestai.] This way for similar cases.
24	*decorative border*

I ↓

- - - - -

1   [   ].[. .]..[   ]
2   [   ]. οἰκῶν ἐν [   ]
3   [c. 7 Ἀ]τρῆτος ἀπὸ κώμη[ς NN τοῦ   ]
4   [c. 4]υ νομοῦ χαίρειν. ὁμολογῶ [   ]
5   [c. 3 ἐ]μαυτοῦ ἀπὸ τιμῆς ʽμναέων' ἐρίου ἥν [.].
6   [c. 5] ειων ἀργυρίων δηναρί[ων   ]
7   [c. 5] ψν. . . . . ἅπερ ἐπαναγκὲς ἀ[ποδώσω   ]
8   [μηνὸ]ς Μεχεὶρ τῷ ἐνεστῶτους ἔτους . [   ]
9   [ἄνευ ὑπ]ερθέσεως καὶ εὐρυσιλογίας, γινομένης [τῆς]
10 [πράξεω]ς ἔκ ται ἐμοῦ καὶ ἐκ τῶν ὑπαρχόντων μ[οι πάντων   ]
11 [κ]ύριον τὸ χιρόγραφον ἁπλοῦν γραφαὶν καὶ ἐπ[ερωτηθεὶς ὡμολό-]
12 [γησα]ʺ.

13 [- 16 -]‾ ἐν δέσαρα μόροια [   ]
14 [- 16 -]′ οβ   *line filler*
15 *decorative border*

16 [   ] ἔ⟨σ⟩ται ἡ ναύβιο⟨ν⟩ [
17 *diagram* θ̄ | . | ιε | .

18 [   ] ἔσται ἡ σπραγί (ἄρουρα?) αδ[   ]
19 *decorative border*

5 *l.* μναείων | 8 *l.* τοῦ ἐνεστῶτος | 9 *l.* εὐρησιλογίας | 10 *l.* τε | ὑπαρχόντων *pap.* | 11 *l.* χειρόγραφον | *l.* γραφὲν | 13 *l.* τέσσαρα μόρια | 14 *l.* οβ′ | 18 *l.* σφραγίς

I ↓

(i3: model contract)

- - - - -

1   [   ]. [. .]. . [   ]
2   [   ]. living in [   ]
3   [c. 7] son of Hatres, from the village of [NN of   ]
4   [c. 4] nome, greetings. I acknowledge [   ]
5   [c. 3] myself from the price of mnai of wool, which [   ]
6   [c. 5] . . . denarii of money [   ]
7   [c. 5] 750 . . . . which I shall of necessity repay [   ]
8   in the month of Mecheir of the present year . [   ]
9   [without] postponement and excuse, with you having [the (right of)]
10   execution upon me and [all] my possessions [
11   The contract, written in a single copy, is authoritative, and having been asked the formal
question, [I gave my assent.]
12

(i4: partition into unit-fractions)
13   [Remainder n, 1/nn] in four unit-fractions [   ]
14   [   ] 1/72.   *line filler*
15   *decorative border*

(i5: mathematical problem)
16   [   ] the trench? will be [
17   *diagram*

```
 \ 9
  |
  | ...               15
  |_____
  |
  | ...
```

18   [   ] the plot will be 1 1/4 arouras?[   ]

III. Commentary

A recto

a1: model contract

The torso of a loan of money, a type also found in the Panopolite codex BL Add. MS 33369. Before damage, the names of the parties (whether "real" names or indefinites such as Τίς) would have occupied the present line 1 and at least one preceding line. In all likelihood the name (in the dative) and legal residence of the lender came first, with the present line 1 containing the name of the borrower in the nominative. Nothing of this can now be recovered with any confidence from the scanty traces. Like the other contracts in the codex, this one lacked the consular date that in a genuine contract would have stood at the start (see, e.g., *SB* 14.12088, which we will cite below as one of the better parallels to this text).[1] Similarly, at the end the recapitulation of the agreement with the subscription of the borrower is omitted, along with any necessary subscription of the person who might have written for an illiterate borrower. Presumably aspiring contract-writers learned the formulas necessary for these parts of the overall contract structure separately.

Loans of money from the Oxyrhynchite nome from the fourth century are not numerous—hardly a dozen, and most of these not fully preserved. It is therefore not surprising that we do not have good parallels for some of the phraseology found in our text. Relevant examples are *P.Amst.* 1.44, *P.Select* 7, *P.Oxy.* 14.1712, 61.4124 and 4125, 72.4895–4897, *P.Oslo* 2.41, *P.Coll. Youtie* 2.82, and *SB* 14.12088, mentioned above. The Hermopolite loan of money *P.Lips.* 1.13 is also an important parallel cited below. For a discussion of fourth-century loans of money, and particularly of interest rates, see *P.Kellis* 1, pp. 115–120. No date is preserved in the body of the text; a year would have been given in line 6 (see note ad loc.). It is interesting that no exactly dated loan in Worp's list dated before 364 is denominated in solidi rather than in the copper

---

1. Such dates are similarly lacking in the Panopolite codex BL Add. MS 33369.

currency. In other contracts of the fourth century, it is only after around 350 that solidi come to be used regularly for prices; similarly, solidi are almost entirely absent from the archaeological record before the middle of the fourth century. For a detailed discussion of the monetary transition in the reign of Constantius see Bagnall and Bransbourg 2019.

2–3. The standard phrasing of loans of money from the Oxyrhynchite in this period includes four key elements at this point; as in the example mentioned above, *SB* 14.12088, we expect παρὰ σοῦ ἐν χρήσει διὰ χειρὸς ἐξ οἴκου σου: "from you, as a loan, from hand to hand, out of your house." The width of the column below, where it is preserved to its full width, shows that the first element stood in line 2 (where pi is readable), and the last (misspelled) stands in line 3. But with only about 8–9 letters available in the lacuna at the start of line 3, we must apparently choose between the two middle elements. In our view, ἐν χρήσει is the more necessary, and we suppose that the writer has omitted διὰ χειρός. The editors of *P.Amst.* 1.44, faced with a similar problem, instead omitted (without comment) ἐν χρήσει. It appears in any case to be impossible to find any place to include a phrase found in almost all loans of money from the fourth-century Oxyrhynchite and Hermopolite, εἰς τὴν ἰδίαν μου καὶ ἀναγκαίαν χρείαν, "for my private and pressing need." This phrase is also lacking in *P.Amst.* 1.44 and not restored there.

3–4. For this combination of adjectives modifying the solidus, see *P.Lips.* 1.13 (Hermopolis, 364), νομισμάτια δεσπ[ο]τικὰ [εὐ]χάρακτα δίζῳδα δύο. δίζῳδον (with image on both sides) as a modifier of the solidus is rare, with the few datable instances found only between 359 (*BGU* 1.316, from Askalon) and 384 (*P.Gen.* 1.12 [2nd ed.]). *P.Ross.Georg.* 3.9 is probably later fifth century, according to J. Gascou (email).

4–5. In all likelihood, ἀντί points to phraseology like that of *P.Oxy.* 45.3266 = *P.Coll.Youtie* 2.82, lines 9–13: ἐπὶ τῷ με ἀντὶ τοῦ αἱροῦντός σοι μέρους τῆς τούτων ἐπικερδείας τελέσιν σοι καθ' ἕκαστον μῆνα [ἀπὸ] τοῦ ἐξῆς μηνὸς Θὼθ τοῦ εἰσιόντος λβ κβ ιδ ε γ (ἔτους) [ἀργυρ]ίου τάλαντα δέκα καὶ τὸ προκίμενον κεφάλαι[ον ἀ]κίνδυνον. But space does not allow anything like as full as that formula, and given the syntax of ll. 5–6 it is not obvious what to restore here.

5. ἀποτάκτου ἐπικερδίας ("fixed interest") occurs to our knowledge otherwise only in *P.Lips.* 1.13, cited in the note to lines 3–4, where ὑπὲρ λόγου ἀποτά[κτο]υ ἐπικερδίας occurs in lines 12–13. In general, ἀπότακτος (like the abstract ἀπότακτον) is used in reference to rent rather than to the yield on a loan. As part of the phrase is lost here, we cannot be sure if in fact the return on the loan was fixed rather than calculated as a proportion per unit of time (as it is in the Leipzig papyrus). ἐπικερδία, which does not appear in the papyri before the reign of Diocletian and is not in general very common, is used in other documents with reference to time-based interest. There are several instances in the Kellis papyri, all fourth century (*P.Kell.* 1.44.8n., citing for the terminology Finckh 1962). See also *SB* 14.12088.11 (ἀντὶ λόγου ἐπικερδίας) and the note in H. C. Youtie's first edition of this papyrus, Youtie 1976, 141, note on line 11.

6. This line apparently forms part of the conditions of the payment of interest on the loan. Since the loan is open-ended (see line 8), this must represent the starting date from which interest is calculated. Plausible restorations would be [ἀπὸ νεομηνία]ς or [ἀπὸ τοῦ ὄντο]ς, requiring respectively 11 and 10 letters. The latter seems more likely, not only for its slight edge in length but because it is more common in late antique Oxyrhynchite contracts. A fourth-century example is *P.Harr.* 1.82 (345), and there are more instances in the fifth century. It should be noted, however, that these are mainly leases, and we cannot cite a direct parallel from a contemporary loan document.

The year originally mentioned at the end of this line is entirely lost, but the presence of the word ἔτος shows that we are dealing with one of the Oxyrhynchite eras based on regnal years. See the note to H ↓.4–5 for these years. After tau at the end of the line, there are the remains of one letter, or possibly two. It is not easy to read the desired omicron of ἔτους, and the descending stroke curves slightly to the right, unlike the bottom of upsilon in this text. Possibly instead ἔτι (*l.* ἔτει) was written.

7. *P.Oxy.* 61.4125.24, in the context of the interest due if repayment of a loan does not occur on time, speaks of τόκον τὸ⟨ν⟩ σταθέντα πρὸς ἀλλήλους (see also *P.Oxy.* 61.4124.16, with note *ad loc.*). The same phrasing occurs in *SB* 14.12088.22–23. We do not think there is enough evidence to restore the lacuna at the start of this line with confidence. It apparently refers back to line 5, where τόν seems to start the phrase describing the amount to be paid as a fixed return on the loan (see note on line 5).

The adverb ἀκοιλάντως occurs in various types of documents, mostly later than the fourth century and more often with reference to payment of rent than of interest or principal of loans.

10. There are only the scantiest traces of the needed sigma at the start of the line. The entire word συ (*l.* σοι) is written to the left of the edge of the text block and may have been added as an afterthought.

11. A single copy of loan documents is not unusual; cf. once again *SB* 14.12088.26. The creditor presumably kept this single copy, returning it to the borrower at the repayment of the loan. But loans in duplicate are also attested, as with *P.Oxy.* 61.4124 and 4125, in both of which two copies on a sheet of papyrus are preserved, still joined.

a2: mathematical problem

Volume problem, trapezoidal solid.
Units: Linear cubits → volume cubits → naubia.
Algorithms: trapezoid area algorithm A (P3A), prism volume algorithm (S2).
Diagram: None.

14. διάκοπος. The term could refer either to an unwanted breach in a dike, as in *P.Oxy.* 16.1409.16, where the dioiketes orders breaches to be "filled up" (ἀποφραγῆναι), so as to contain the water from the inundation, or to an intentional breach to allow water to reach fields. See *P.Petaus* 18.24–25n. for the varying usage. The problem is to calculate the volume of earth equivalent to the breach. The student was expected to recognize that a dike is a trapezoidal prism of indefinite length, while the breach is a section cut off from the dike by two planes perpendicular to the prism's axis (Fig. 24).

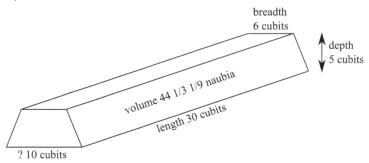

breadth
6 cubits

depth
5 cubits

volume 44 1/3 1/9 naubia

length 30 cubits

? 10 cubits

Fig. 24. Reconstructed diagram for problem a2.

15–16. The calculation shows that 10 cubits and 6 cubits are the parallel (horizontal) sides of the trapezoidal cross section, 5 cubits is the altitude, 30 cubits is the perpendicular dimension of the prism, i.e., measured along the dike. The shorter (upper) horizontal dimension of the cross section is called "width." The word at the ends of lines 15 and 17 is unclear, but is neuter and begins with σαν- (as is clear from line 17); since it is added to the upper width given in line 16, and the sum is then divided by two, it must refer to the lower width.

20. 1200/27 = 44 4/9, correctly expressed as 44 1/3 1/9.

A verso

a3: mathematical problem

Volume problem, trapezoidal solid.
Units: Linear cubits and fingers → cubits × fingers^2 → unnamed volumetric units and solid fingers.
Algorithms: trapezoid area algorithm A (P3A), prism volume algorithm (S2).
Diagram: seemingly unrelated to the problem (see below).

The initial specification of the solid, as well as one of the dimensions, are lost; there was probably a single line of text above line 1. The calculation implies that it is a trapezoidal prism, and that the parallel sides of the cross section have lengths adding to 14 of some unit, while the altitude (here, "width") is 8 of some unit, and the perpendicular dimension (here, "thickness") of the solid is 4 fingers. The dimension given as 4 fingers at the beginning of line 2 is thus probably the "thickness." The 6 cubits at the beginning of line 1 must be one of the parallel sides, so that the other parallel side (presumably given in the lost preceding line) should have been 8 cubits. The difference between the two parallel dimensions of the trapezoid, 2 cubits, is very large relative to the altitude (8 fingers = 1/3 cubit) and the perpendicular dimension of the solid (4 fingers = 1/6 cubit); this was effectively a long squared rod with tapered ends (see Fig. 25). It is difficult to imagine the kind of object that was supposed to have this shape.

The volume calculated at the end of line 4, namely 224, is in units of one cubit by one finger by one finger, i.e., 1/576 of a cubic cubit. The division by 288 in line 5 amounts to a conversion to an unnamed unit equivalent to half a cubic cubit, which is further subdivided into 24 volu-

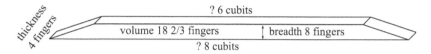

Fig. 25. Reconstructed diagram for problem a3.

metric units called "fingers." This unnamed unit is also employed in problems c1 and g4. In problem b5 an unnamed volumetric unit arising in a similar manner is equivalent to one third of a cubic cubit, though it still comprises 24 volumetric "fingers."

The badly damaged diagram following this problem appears to show two concentric circles, with textual labels to their left indicating a circumference in cubits, a "width" in fingers (again no numeral), and a "thickness" in fingers (?), all apparently lacking numerals. The number 18 appears by itself to the right of the drawing. Except for the fact that the unit of "thickness" is fingers, none of the information in the diagram matches that in the problem (unless the 18 represents the whole-number part of the solution), so it seems to be an erroneous intrusion.

a4: partition into unit-fractions

See introduction, section 10 for problems of this type and a reconstruction of a possible method of solution.

9. The four unit-fractions as given do not sum to 4 ÷ 80. Assuming that only one of them is wrong, the only possible correction is 91 for 95. This is more likely a copying error than one of calculation.

a5: mathematical problem

Area problem, trapezoid.
Units: schoinia → arourai (not explicitly named).
Algorithm: trapezoid area algorithm B (P3B), incompletely applied; inverse diagonal rule (P1i).
Diagram: trapezoid divided into two right triangles flanking a rectangle, labelled with numerals for linear and area dimensions but no verbal text.

The problem (Fig. 26) is abstract except for the specification of linear units for the oblique sides ("legs") in lines 11–12. All the dimensions, however, must be in the same units for the inverse diagonal rule to be valid. The solution, involving an application of this rule to obtain the altitude of the trapezoid, is reminiscent of the problems in *P.Chic.* 3 and *P.Bagnall* 35, but the equality of the oblique sides here makes the algorithm simpler. The dimensions have been chosen so that the trapezoid is subdivided into two 3–4–5 triangles flanking a rectangle.

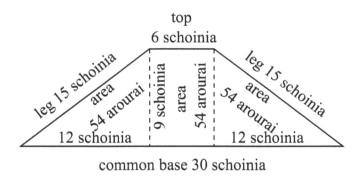

Fig. 26. Reconstructed diagram for problem a5.

11. τραπάρδιον is presumably a garbling (not a mere phonetic distortion) of τραπέζιον.

12–13. ἐπειδὴ is unexpected.

18. In *P.Bagnall* 35 the altitude of the trapezoid is called ὀρθὴ βάσις, "upright base." It is the altitude of the right triangles as well as of the rectangle.

19–20. These steps lead to the area of either of the two equal right triangles.

B recto 1–2. To complete the problem, one should add the area of the rectangle and the areas of the two equal right triangles, thus $A = (54 + 54 + 54)$ schoinia^2 = 162 schoinia^2. Of course the doubling of the triangular area cancels the division by 2 in the calculation of that area, so the procedure could have been made more efficient at some cost in didactic clarity.

The numerals in the diagram are placed in appropriate positions, except that the upper (shorter) parallel side is labelled "15." The meaning of the numbers, as linear dimensions (15 twice, 12, 9), as areas of parts of the figure (54 twice), or as squares of linear dimensions (144, 225), would only be recognizable through reading the text.

B recto (continued)

b2: mathematical problem

Area problem, square.
Units: schoinia → arourai (not explicitly named).
Algorithm: rectangle area algorithm (P2).
Diagram: None.

The fragmentary text at the end of line 5 looks similar to the use of forms of ὁσαδήποτε to mark an unknown quantity in problems d4, e2, and o2, but the solution shows that the unknown here is the field's area, while its dimension from south to north is, like that from west to east, 4 schoinia, i.e., it is a square (Fig. 27). The problem is trivial, requiring only the algorithm for the area of a rectangle with no unit conversions.

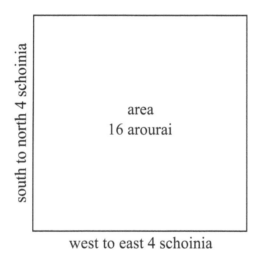

Fig. 27. Reconstructed diagram for problem b2.

2. Small sloping and elevated tick marks, single or multiple as here, are frequent in the manuscript not only after numerals but also after words and phrases, sometimes apparently marking items in a list but often serving no obvious purpose. For a variant form see the commentary to E verso line 16.

b3: mathematical problem

Volume problem, complex shape.
Units: cubits → volumetric cubits → bricks.
Algorithms: hollow rectangle area algorithm (P5), prism volume algorithm (S2), unidentified algorithm.

Diagram (on B verso): damaged.

The volumes of two components of the complete structure are calculated in volumetric cubits, added together in 19–20, and the sum converted to bricks using a coefficient of 48 bricks per volumetric cubit. The first component, presumably the tower proper, is to be understood as a rectangular wall 2 cubits thick, and having its exterior perimeter 20 cubits—which would make it rather tiny!—and its interior perimeter 18 cubits. This is an impossible configuration, because the thickness of the wall must be one quarter of the difference between the perimeters, thus here 1/2 cubit.[2] The person who devised the problem must not have realized that the thickness is determined by the two perimeters.

15–17. The calculation of the volume of the tower's walls is performed similarly to the calculation of the volume of a prism having a trapezoidal cross-section, with the outer and inner perimeters taking the place of the longer and shorter parallel sides of the trapezoid, and the wall's thickness taking the place of the trapezoid's altitude. This algorithm is justifiable since, whatever the specific length and breadth of the tower are, the horizontal cross-section of the wall can be dissected into four trapezoids of equal altitude.

17–19. We suspect that the substructure was supposed to be understood as a rectangular platform of bricks beneath the tower, so that the calculation should have involved multiplying the given length (10 cubits) by the given breadth (8 cubits) and then by some thickness which was left unstated. Instead, the solution offered treats this component of the structure as a pair of walls of lengths 10 and 8 cubits and having the same thickness and height as the walls of the tower. It is not clear how such an annex to the tower might have been visualized; the incompletely preserved diagram at the top of B verso is, as usual for this manuscript, no help.

20. ἐπὶ τὸν $\overline{\mu\eta}$. The coefficient 48 bricks to a volume cubit must have been standard in mathematical problems since it appears also in the two sixth-century papyri *P.Lond.* 5.1718.78, and BM Add. MS 41203A, recto line 5 (edited in Skeat 1936, with discussion of this point at 21–22 and corrected reading of the *P.Lond.* passage at 20). In reality, brick dimensions were quite variable in the Roman period, as in all periods of Egyptian history; see Spencer 1979, 109–110 (for bricks in fortifications) and Plate 42. For example, contemporary with *P.Math.* are domestic and military buildings at Amheida, Dakhla Oasis and in the northern Kharga Oasis, constructed with bricks of approximately 33–34 cm × 16–17 cm × 7–8 cm, which (neglecting mortar) would yield a coefficient of about 35 bricks to the cubic cubit; see Boozer 2015, 143–144 and Rossi & Fiorillo 2018, 378–382. A brick size 30 cm × 15 cm × 6 cm, however, is well within the range of common Roman-period sizes, and would give about 54 bricks to the volume cubit, neglecting mortar, but even small variations in the linear dimensions would change the number so as to

---

2. Assuming a different outline for the tower instead of a rectangle would result in a different thickness, but under no circumstances could the thickness be equal to the difference between the perimeters.

make a standard coefficient inaccurate in practice. As a rough rule of thumb for a builder who needed to order bricks to be made, however, it might be close enough to be adequate. And of course it could be replaced by another coefficient if one knew the module to be used.

B verso

b4: mathematical problem

Area problem: rectangle.
Units: schoinia → arourai (not explicitly named).
Algorithm: See commentary.
Diagram: apparently a right-angled triangle haphazardly labelled with the various numerals occurring in the problem and solution.

The area should, of course, be computed by multiplying the two dimensions together (Fig. 28). Instead, one side is squared, then multiplied by the other side, and lastly divided by the first side, a perversely roundabout route to a correct result.

Fig. 28. Reconstructed diagram for problem b4.

4. μ[. We expect ποιοῦμεν, but the μ is clear and the available space small. Could the writer have accidentally omitted the first syllables?

b5: mathematical problem

Volume problem, trapezoidal prism.
Units: cubits and fingers → cubits × fingers² → unnamed volumetric units and solid fingers.
Algorithm: trapezoid area algorithm A (P4a), prism volume algorithm (S2).
Diagram: schematic drawing suggesting a chest shape with a rounded (cylindrical)) top, seen from above, labelled haphazardly with some of the numerals occurring in the problem and solution.

The solid (Fig. 29) is described rather obscurely as a "quadrangular trapezoid," but the solution implies that it is again a trapezoidal prism. The parallel sides of the trapezoidal cross section are called the "breadth" and the "top," the altitude is called the "thickness," and the perpendicular dimension of the prism is the "length." Although the "breadth" and "top" are stated as being in different units, respectively cubits and fingers, the two associated numbers are added together with no unit conversion as the first step of the calculation of the cross section's area. Apparently "cubits" is here a mistake, and both measurements are actually fingers. Hence the volume obtained at the beginning of line 14 is in units of one cubit by one finger by one finger, as in problem a3. Dividing by 192 converts to a volumetric unit equal to one third of a cubic cubit; this is further subdivided in to 24 volumetric fingers. It is probably pointless to speculate on what kind of rod-like object would have a length of 48 cubits.

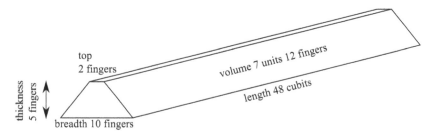

Fig. 29. Reconstructed diagram for problem b5.

C recto

c1: mathematical problem

Volume problem, quadrangular prism.
Units: cubits and fingers → cubits × fingers² → unnamed volumetric units and solid fingers.
Algorithm: quadrilateral area algorithm (P4), prism volume algorithm (S2).
Diagram: schematic drawing suggestive of a trapezoidal prism seen from above, labelled with a few of the numerals occurring in the problem.
In the solution, each pair of opposite sides is designated by a single name, "width" and "thick-

ness," and all are presumably expressed in fingers, while the perpendicular dimension of the prism, the "length," is in cubits (Fig. 30). In line 4, however, one of the "thickness" dimensions

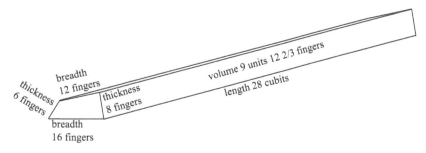

Fig. 30. Reconstructed diagram for problem c1.

seems to be referred to by an allusion to "leaves" (φύλλων), unless this is a phonetic distortion of a word we have not recognized. We can make no sense of the other damaged names for the sides in lines 2–3. The unnamed volumetric unit of the final result is the same as in problem a3.

c2: mathematical problem

Fractions and proportion.
Units: artabas.
Algorithm: scaling of fractions.
Diagram: None.

Although line 14 appears to speak of *loading* a boat with a cargo of wheat, the subsequent information only makes sense in terms of a partial *unloading*. The text is otherwise mostly straightforward; *P.Cair. cat.* 10758, problem 13 is of the same type.

18–19. Once the fractions have been added together, the next step is to determine what supplement of unit-fractions has to be added to the sum to make a complete 1 (here called "the monad"); for parallels that also show how such completions were carried out, see *P.Cair. cat.* 10758, problems 14, 15, and 32. Here the operation is expressed as a question and answer, the precise expression of which is obscure. We believe that the bizarre-looking text is best explained as representing the question, "how much is the whole number 1 missing?," i.e., the goal is to account for one complete boatload, and we have so far 1/2 + 1/3 + 1/12 of the load, so what fraction of the load is still missing? The indicative question τί λείπει ἡ μονὰς μία; somehow assumed the infinitive form τί λείπειν τὴν μονάδα μίαν; before undergoing further linguistic distortion.

C verso

c3: mathematical problem

Side and area problem: isosceles triangle.
Units: schoinia → arourai (not explicitly named).
Algorithm: isosceles triangle sides algorithm A (P8A), triangle area algorithm (P6).
Diagram: a right-angled triangle crossed by a vertical line, haphazardly labelled with some of the numbers occurring in the problem and solution.

The statement of the problem appears to be identical to the duplicate problems d1 and g3, where the givens are the lengths of the base and of the equal sides of the triangle, while the first unknown to be found is the triangle's altitude (Fig. 31). ἀνά is normally the signal for the application of a number or series of numbers to each of a set of dimensions. This cannot be the interpretation in the present instance, however, since the base, 48 schoinia, would be longer than the sum of two equal sides of 18 schoinia. The solver has thus correctly understood the problem to be asking for the length of the equal sides on the assumption that the *altitude* is 18 schoinia. But in the second part of the solution, the solver gets confused and takes this calculated length of the equal sides as if it was the altitude; the correct solution would be 864 = 48 × 18.

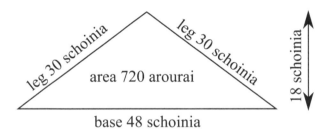

Fig. 31. Reconstructed diagram for problem c3.

c4: partition into unit-fractions

See introduction, section 10 for problems of this type and a reconstruction of a possible method of solution.

c5: mathematical problem

Volume problem, rectangular parallelepiped.
Units: schoinia, xyla, cubits → cubits → volume cubits → naubia.
Algorithm: parallelepiped volume algorithm S1.
Diagram: schematic figure suggestive of a triangular prism, labelled haphazardly with some of the numbers from the problem and solution.

13. διῶραξ ποταμοῦ is paralleled in Pierpont Morgan tablet 3 verso lines 1–2 διῶρυξ ποταμὸν (the parallel argues against punctuating between the words; perhaps one should read ποταμῶν?). *P.Cair. cat.* 10758, problem 5 has διόρυξ τετράγωνον; *P.Lond.* 5.1718.74 has simply διώρυγα μετρῆσαι. We suggest that the ποταμός in question is meant to be a canal, not a river, even if the given width is unrealistic. In all these texts, the object of the problem is evidently some kind of artificial trench, assumed to have vertical sides and a horizontal bottom (Fig. 32). The main point of the exercise seems to be the unit conversions.

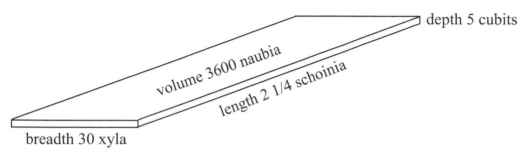

Fig. 32. Reconstructed diagram for problem c5.

D recto

d1: mathematical problem

Altitude and area problem: isosceles triangle.
Units: schoinia → arourai (explicitly).
Algorithm: isosceles triangle altitude algorithm B (P8B), triangle area algorithm (P6).
Diagram: similar to diagram for c3.

Problem g3 is a duplicate of this problem. In both versions there is a fragmentarily preserved first line whose function is hard to explain since the statement of the problem appears to begin in normal fashion in the second line. After that point, aside from orthographic variants, the texts are practically identical. Significant variants: d1 line 4 ποιοῦμαι / g3 line 3 ποιοῦμεν; d1 lines 4–5 βάσεως (apparently, from available space) / g3 line 4 κοινῆς βάσεως (apparently, from available space); d1 line 12 (diagram) αὐτὴ / g3 line 9 (diagram) ἑκάστη ὀρ(θὴ). Even recognizing the formulaic style of these problems, it is hard not to conclude that a common written source lies behind both versions.

The statement of the problem is the same as in c3 except for the choice of numbers, but here, notwithstanding the injunction in line 3 "to find the other sides," all sides have been given at the outset and the unknowns are the triangle's *altitude* and area (Fig. 33).

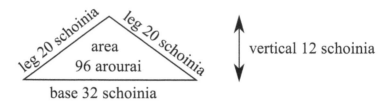

Fig. 33. Reconstructed diagram for problem d1.

d2: mathematical problem

Arithmetical problem: times, speeds, distances.
Units: stades.
Algorithm: see below.
Diagram: none.

The statement of the problem is rather elliptical but the scenario can be filled out on the basis of the solution. Runner *A*, who runs at a speed of 9 stades per day, sets out on day 1. Twelve days later, on day 13, runner *B* sets out in pursuit of *A* at a speed of 15 stades per day; meanwhile it is tacitly assumed that *A* continues at his own speed. The question is to find how many days, counting from day 13, it will take for *B* to catch up with *A*.

16–17. The first step of the solution is to calculate the distance that *A* has already travelled at the point when *B* sets out, which is 108 stades.

17. Next, the relative speed of *B* with respect to *A* is found as 6 stades per day. This is the rate at which the distance between *B* and *A* diminishes.

18–19. Lastly, one divides 108 by 6 to obtain the problem's solution. Unusually, the text does this by finding what fraction 6 is of 108 rather than by calling for a straightforward division.

D verso

d3: mathematical problem

Sides and area problem: right-angled triangle.
Units: schoinia (not explicitly named in extant text) — arourai.
Algorithm: scaling, right-angled triangle area algorithm (P7).
Diagram: right-angled triangle labelled with some of the numbers occurring in the problem and solution.

The problem (Fig. 34) has three components: (i) to find the sides of a "Pythagorean" triangle with sides in the ratio 3 : 4 : 5 such that the sum of the sides plus the perimeter (which is of course just the sum of the sides again!) equals 192; (ii) to find the area of the triangle; and (iii) to find the perimeter and the sum of the sides plus the perimeter. This last result might charitably be thought of as a check of the correctness of the preceding calculations, though nothing in the text signals that this section has such a function.

We may remark on two peculiarities of this problem: the apparently contrived given quantity and the given fact, brought explicitly to the reader's attention, that the triangle in question is a "Pythagorean" right triangle. The designation "Pythagorean" either means, more generally, that this is a right triangle with sides that are all integer numbers of some common measure, or, more specifically, that it is the familiar right triangle with sides in the ratio 3 : 4 : 5, in which case the text's specifying the 3 : 4 : 5 proportionality is redundant.[3] Part (i) uses this proportionality, but not the fact that the triangle is right-angled, so that any proportionality, i.e., any shape of triangle, could have been stipulated. Part (ii) depends on the fact that it is a right triangle, but any right triangle would work here since no use is made of the fact that the hypotenuse is a whole number of units. Hence while it is true that the exact solubility of the problem *taken as a whole* by the methods employed in the text depends on being given that the triangle is right-angled and has whole-number sides, the condition does not lead to mathematically interesting consequences. In particular, the diagonal rule and inverse diagonal rule are not invoked.

Our problem's oddities become explicable when it is compared with certain problems preserved in other sources:[4]

(*P.Gen.* 3.124)
(i) Given the lengths of one leg and the hypotenuse of a right-angled triangle, find its other leg.
(ii) Given the length of one leg and the sum of the lengths of the hypotenuse and the second leg of a right-angled triangle, to find the hypotenuse and the second leg separately.
(iii) Given the length of the hypotenuse and the sum of the lengths of the two legs of a right-angled triangle, to find the two legs separately.

(*Corpus agrimensorum,* "Marci Iuni Nipsi Podismus")[5]
(iv) Given the area and the length of the hypotenuse of a right-angled triangle, to find the two legs.
(v) Given the area, the length of the hypotenuse, and the sum of the lengths of the legs of a right-angled triangle, to find the two legs separately. (This problem is overdetermined.)

---

3. Cf. Iamblichus, *Theologoumena arithmeticae* p. 50 line 21 (ed. de Falco), Πυθαγορικῷ ὀρθογωνίῳ τριγώνῳ meaning the 3–4–5 triangle. For γ δ ε as a way of designating the 3–4–5 triangle, see pseudo-Heron, *Geometrica* 24.4 line 4, *Geodaesia* 12.2 lines 1–2, Diophantus, *Arithmetica* ed. Tannery p. 368 l. 17 etc., and Proclus *In Platonis rem publicam comm.* v. 2 p. 40 l. 28.
4. See Sesiano 1999, with references to the parallels in the Corpus Agrimensorum and Heron on p. 27 n. 5.
5. Blume, Lachmann, and Rudorff 1848–1852, 1.297–299.

(pseudo-Heron, *Geometrica* ed. Heiberg, pp. 422–426)
(vi) Given the sum of the perimeter and the area of a right-angled triangle (assuming homogeneity of linear and area units!), to find the three sides and the area separately. (This problem is indeterminate.)

The problems in the Geneva papyrus, in particular, are clearly part of a sequence of related problems, at least one of which must have preceded (i) since this section begins ἔστω δὲ πάλιν τρίγωνον. Problem (i) requires a simple application of the inverse diagonal rule, while the others are more sophisticated problems solved by algorithms based on a quasi-algebraic analysis of the configuration in which implicit use is made of the Diagonal Rule and the sum-and-difference rule. The triangles in such problems *must* have whole-number sides if the solutions are to be exact, and accordingly the specific triangles employed in the texts are the 3–4–5, the 5–12–13, and the 8–15–17 triangles. Our d3, the conditions of which have a particular resemblance to (vi), thus appears to be an elementary problem selected from—or debased from—a repertoire of problems involving varying combinations of dimensions of right-angled triangles.

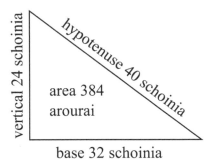

Fig. 34. Reconstructed diagram for problem d3.

5. ἀπόδιξεις (*l.* ἀπόδειξις) occurs in our codex here and in two other mathematical problems (see f1, F recto 2 and o5, O verso 7) as a header-word preceding part of the problem's treatment. In other mathematical papyri, e.g., *P.Mich.* 3.142 (i 4 and elsewhere), it signals that what follows is a verification that a solution obtained in the preceding text satisfies the conditions of the problem; for full references see Liesker and Sijpesteijn 1996, 184). On the other hand, in the papyrus (re-)edited in that article, *MPER* NS 15.172–174 (= *SB* 24.16273), lines 1 and 3, ἀπόδειξις precedes what appear to be the actual solutions of two conjoint problems, to find the diameter and area of a circle of circumference 30 schoinia. If we understand the present problem and o5 correctly, the word intervenes between steps of the solution, perhaps indicating the beginning of a new stage of the operations; in f2 the usage is unclear because of the poor state of preservation of the problem.

d4: mathematical problem

Linear dimension from volume problem: cylinder.
Units: naubia, cubits → volume cubits → cubits (not explicitly stated).
Algorithm: inverse prism volume algorithm (S2i), inverse circle area algorithm A (P10Ai).
Diagram: none.

Although the statement of the problem speaks of the "upper diameter," no reference is made to the base of the excavation, and the data have evidently been chosen on the assumption that the solid in question (Fig. 35) is a cylinder (likewise in problems e1 and e2). Problem e1 concerns the identical situation but with a different disposition of knowns and unknowns, while e2 concerns a similar situation but with different dimensions as well as a different disposition of knowns and unknowns.

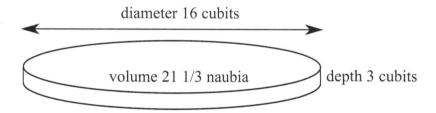

Fig. 35. Reconstructed diagram for problem d4.

E recto

e1: mathematical problem

Volume problem: cylinder.
Units: cubits → volume cubits → naubia.
Algorithm: circle area algorithm A (P10A), prism volume algorithm (S2).
Diagram: circle and diameter, labelled with some of the numbers occurring in the problem and solution.

This problem (Fig. 36) is a permutation of problem d4. Perhaps the close relation between them explains the absence of a decorative border separating them.

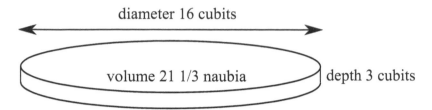

diameter 16 cubits

volume 21 1/3 naubia   depth 3 cubits

Fig. 36. Reconstructed diagram for problem e1.

e2: mathematical problem

Linear dimension from area problem: cylinder.
Units: naubia, cubits → volume cubits → cubits.
Algorithm: inverse prism volume algorithm (S2i), inverse circle area algorithm B (P10Bi).
Diagram: circle, labelled with some of the numbers occurring in the problem and solution.

For similar problems see d4, e1.

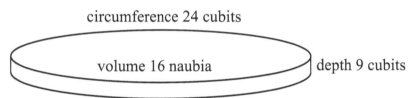

circumference 24 cubits

volume 16 naubia   depth 9 cubits

Fig. 37. Reconstructed diagram for problem e2.

E verso

Our transcription of this page is supplemented at the beginnings of the lines by readings from an early 1980s photograph that shows the page before breakage removed much of its left edge (see Introduction, section 1). Letters visible in the photograph but no longer extant are printed in italics.

e3: metrological text

Lines 1–15 have an extensive parallel in *P.Oxy* 4.669 col. i (col. ii parallels our text g1). *P.Oxy* 4.669.i.10ff has a quasi-tabular format, with the numerals lined up in a column towards the right of the text column. Lines 14–18 are in part paralleled, with minor verbal variants, by text m1 lines 10–13.

1–4 = *P.Oxy* 4.669.i.1–4. Relations of linear units: γεωμετρικὸν σχοινίον (surveyor's schoinion), ὄγδοον (ogdoon), πῆχυς (cubit), and εὐθυμετρικὸν σχοινίον (linear schoinion). The nonorthographical variants between the two texts are slight: l. 2 εἶναι = *P.Oxy* 4.669.i.2 ἔχειν, l. 3 ἐστιν om. *P.Oxy* 4.669.i.3

4–6 = *P.Oxy* 4.669.i.5–8 (possibly 5–9). Definitions of linear, area, and volume cubits. In our papyrus, probably owing to eyeskip, the definition of the area cubit and the name of the volume cubit are omitted after line 5 so that three dimensions are ascribed to the area cubit.

6 βάθος ἥδε ὕψο[ς] ἢ πάχος: *P.Oxy* 4.669.i.8 has βάθος ἤται ὕψος followed, in the next line, by a lacuna estimated at 13 letters, which likely contained ἢ πάχος. ἤτοι is the plausible suggestion of the editors of *P.Oxy* 4.669.

7–8 = *P.Oxy* 4.669.i.9–10. Definition of οἰκοπεδικὸς πῆχυς in terms of area cubits. If ἢ πάχος was written at the beginning of *P.Oxy* 4.669.i.9, the word or words preceding ⟨ο⟩ἰκοπεδικὸς there would have an estimated 9 letters whereas we estimate five letters (one lost, four in very slight traces) in the corresponding place at the beginning of l. 7.

8–13 = *P.Oxy* 4.669.i.11–20. Definitions of βασιλικὸν ξύλον (royal xylon) and ἰδιωτικὸν ξύλον (private xylon) and their relations to the cubit, palm, finger, and surveyor's schoinion.

13–16 = *P.Oxy* 4.669.i.21–25. Definitions of δημόσιον ναύβιον (civil naubion) and ἰδιωτικὸν ναύβιον (private naubion) as cubic xyla. The figures 27 solid cubits and 18 1/2 1/3 1/9 1/54 solid cubits are the cubes of 3 cubits and 2 2/3 cubits, the lengths of the two kinds of xyla just defined. The parallel in text m1 omits the definition of the private naubion; *P.Oxy* 4.669 cuts off just before this point.

16. This is the first of many instances in our manuscript of sloping tick marks accompanied by dots above and below, like a modern percent sign and represented as such in our transcription. We do not believe that these marks signify anything further than the simple tick marks, for which see the commentary to B recto line 2.

16–18. Definitions of artaba and metretes in terms of volume cubits. The versions here in line 18 and in m1 line 13 share the omission of the numeral 3 for the number of metretai in a volume cubit, implying a not too distant common source.

18–24. Definitions of κατὰ πόλιν ἄρουρα (civic aroura), κατ᾽ ἄγρον ἄρουρα (country aroura), and βῖκος, and their relation to area cubits. Apparently the unit ἐν οἰκοπέδοις ἄρουρα (aroura in building lots) named in line 18 is just an alternative name for the civic aroura; since the city area will have been divided into building lots, that is logical. The equivalences of the two kinds of aroura in bikoi are also stated in m1 lines 13–14.

23. ϡϟβ (992) is a copying error for ρϟβ (192). The subsequent calculation is made with the correct number.

F recto

f1: mathematical problem

Arithmetical problem.
Units: artabas, talents.
Algorithm: ?
Diagram: small table.

The nature of this problem is unclear.

f2: mathematical problem

Arithmetical problem.
Units: artabas.
Algorithm: see below.
Diagram: table.

The problem has very close parallels in *P.Cair. cat.* 10758, problems 47–49, which help to clarify severe obscurities in the present example. Each problem begins with a statement that there are three granaries, each containing a specified number of artabas; the numbers range from 200 up to 950. Someone comes along and combines them (μίξας τις) and then takes away (εἶρκεν [*l.* ἦρκεν?] *P.Cair. cat.* 10758, εὑρεῖν here line 5) a specified number of artabas. One has to determine how many artabas came from each granary (e.g., πόσας εἶρκεν [ἀφ'] ἑκάστου, *P.Cair. cat.* 10758, problem 47), on the unstated assumption that their contributions were prorated.

We may take *P.Cair. cat.* 10758, problem 47 as an illustration of the somewhat roundabout methods of solution in that manuscript. There, the three granaries contain respectively 200, 300, and 500 artabas, and a total of 60 are taken away. (i) One first finds the sum of the contents of the granaries, 1000 artabas. (ii) One factors this sum into 10 times 100 (a factorization tacitly chosen such that 100 is also a factor of the amounts in the individual granaries). (iii) Each granary's artabas are then divided by 100 to obtain respectively 2, 3, and 5. (iv) These quotients add up to 10. (v) The 60 artabas are now divided by 10, the other factor, yielding 6. (vi) Lastly, 6 times 2, 3, and 5 results in 12, 18, and 30 artabas as the respective contributions of the three granaries.

From line 12 συντίθω on, the solution in our codex gives steps (iv) through (vi), concluding with a confirmation that the resulting quantities add up to the total artabas that were taken away. The earlier steps, however, are hopelessly garbled and effectively untranslatable. At the outset, in steps (i) and (ii), the sum of the granaries' contents, 900 (ϡ) artabas, ought to have been found, and then 900 should have been factored into 9 times 100. These steps seem to have been skipped over entirely. One suspects, however, that the instances of ἕκαστος/ἑκάστης in

11–12 are mishearings or misreadings of forms of ἑκατοστός, "hundredth" (familiar vocabulary from its use in taxation, using a feminine form) or ἑκατόν, 100 (the word expected in this context). If so, we can conjecture something like the following ideal text, constituting step (iii), as lying behind the mess we read in lines 10–12:

10 οὕτω ποιοῦμεν. σ̄ πό[σα]
11 ἑκατὸν ἔχει; β̄. τριακόσια πόσα ἑκατὸν ἔχει; [γ̄.]
12 ῡ πόσα ἑκατὸν ἔχει; δ̄.

10 We proceed as follows. 200 contains how many
11 hundreds? 2. Three hundred contains how many hundreds? [3.]
12 400 contains how many hundreds? 4.

15–16. The final addition of 140, 210, and 180 constitutes a check that the solution is correct.

f3: partition into unit-fractions

See introduction, section 10 for problems of this type and a reconstruction of a possible method of solution.

21. As given, the unit-fractions do not sum to 9 ÷ 119. Assuming that only one of them is wrong, the only possible correction is 84 for 85.

f4: partition into unit-fractions

See introduction, section 10 for problems of this type and a reconstruction of a possible method of solution.

22. λβ (32) makes no sense here; how it intruded is unclear.

F verso

f5: mathematical problem

Area problem: quadrilateral.
Units: paces, reeds? → cubits → arourai.
Algorithm: quadrilateral area algorithm (P4).
Diagram: none.

It is not clear whether any lines of text are lost at the top.

1–2. The east side is stated to be 15 cubits, and the unit of the west side's 13 units should be the same since the two numbers are added together in line 6. The unit conversion in line 7 shows that this unit is not the cubit after all, but the reed (equivalent to 6 cubits). The quadrilateral thus has an east side 90 cubits long, which is equal to the sum of its west and north sides, respectively 78 and 12 cubits. The person who devised the problem probably did not try to imagine what such a quadrilateral would look like.

8–9. The product of the two averages of opposite sides, 1176, is mistakenly called a "difference." For the equation of 9216 area cubits with 1 (country) aroura see e3, E verso 23–24. Division of 1176 by 9216 yields the quotient 1/8 + 1/384. Instead of the unit-fraction 1/8 (η′), however, the manuscript in both lines 8 and 9 has an odd symbol like a delta with a vertical stroke descending from its lower horizontal stroke, and in line 8 also having a hook at its top—somewhat resembling the monogram for διπλᾶ.

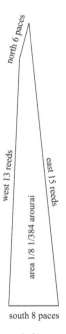

north 6 paces

west 13 reeds

east 15 reeds

area 1/8 1/384 arourai

south 8 paces

Fig. 38. Reconstructed diagram for problem f5.

f6: mathematical problem

Linear dimension problem: right-angled triangle.
Units: none.
Algorithm: unclear; cf. diagonal rule (P1) and inverse diagonal rule (P1i).
Diagram: right-angled triangle labelled with names and lengths of the sides and length of the perimeter.

The problem as stated, to find the lengths of the legs of a right-angled triangle of given hypotenuse, is indeterminate without further data (Fig. 39). One gets the impression that the integral solution offered is supposed to have been discovered by trial and error, rather than by recognizing the given hypotenuse from a list of "Pythagorean" triangles.

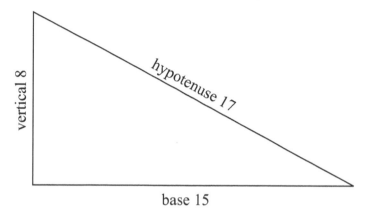

Fig. 39. Reconstructed diagram for problem f6.

f7: mathematical problem

Area problem: rectangle.
Units: cubits → area cubits → vines.
Algorithm: rectangle area algorithm P2.
Diagram: none.

The problem (Fig. 40) is mathematically trivial. A coefficient of one vine to four area cubits, i.e., an area a little more than one square meter, is tacitly assumed.

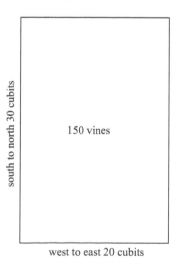

Fig. 40. Reconstructed diagram for problem f7.

G recto

g1: metrological text

Lines 1–5, 10–11, and 15–19 are paralleled in *P.Oxy* 4.669.ii.26–36. Pseudo-Heron, *Geometrica*, ed. Heiberg p. 188 is another parallel for 1–5.

1–5. The listed units from the beginning through schoinos are units of length in order of increasing size.

4. σχοινίον: the editors of *P.Oxy* 4.669.ii.29 restore ἰούγερον in place of this in a lacuna. δίαυλον: after this, *P.Oxy* 4.669.ii.30 has μίλιον, followed by an uncertainly read delta and a lost half line.
   For δόλιχος as a unit of distance see pseudo-Heron, *Geometrica* 4.13 (ed. Heiberg 194), where it is equated to 12 stades or 4800 cubits.

5–6. Line 5, from σχοῖνος, through line 6, ἄμφοδος, is not in *P.Oxy* 4.669 and reads like a compression of m1, M recto 4–8, omitting the metrological relations but senselessly retaining the verb ἐστίν (which we translate "they are," as the best meaning one can give it in this garbled context). For the difficulties involved in the units dodekatikon, laura, and amphodon see the notes to m1. Why the fathom is repeated at the end of 6 escapes us.

6–9. The discussion of the δάκτυλος as the smallest and fundamental unit seems to have a loose parallel in the fragmentary *P.Oxy* 4.669.ii.43–45. It would seem to imply that all the foregoing units were supposed to be of length.

11–19. *P.Oxy* 4.669.ii.32ff consistently omits παλαισταὶ following the numbers.

12. The end of *P.Oxy* 4.669.ii.32 can perhaps be restored οἱ δ ποὺς Α[ἰγύπτιος, οἱ δὲ]. The version in *P.Oxy* 4.669 omits the other varieties of foot.

16. The lost portion of *P.Oxy* 4.669.ii.33 can probably be restored identically to the corresponding part of line 16.

16–17. The text in *P.Oxy* 4.669.ii.34–35 is terser: οἱ ϛ παλεσταὶ [πῆχυς δημό]|σιος κὲ τεκτονικός.

19. The name "loom cubit" (ἰστονικός or conceivably ἰστωνικός) seems not to be attested elsewhere; its name would have fallen in a lacuna at the end of *P.Oxy* 4.669.ii.36. In *P.Oxy* 4.669.ii.37, following right after the 8-palm cubit, the next unit is the βῆμα (pace), stated to be equal to 10 palms and to be the distance between a person's feet. In our text, it appears that a unit equal to 9 palms is the distance between the "legs." Very likely this too is the βῆμα, and the copyist

wrote the wrong number of palms because he expected the number to be one more than the preceding unit, as it always has been up to this point.

g2: mathematical problem

Volume problem: cylinder.
Units: cubits → volume cubits.
Algorithm: circle area algorithm B (P10B) erroneously applied, prism volume algorithm (S2).
Diagram: circle with labels (apparently garbled).

22. The cylindrical object (Fig. 41) is called a "naubion," a word that normally refers to a volumetric unit (which is *not* employed in this problem). The word is, however, similarly used to refer to a parallelepipedal object in mathematical problems in *T. Varie* 71 verso 15, 72 recto 6 and 73 recto 8 (ναύβ(ιον) τετράκ(ωνον?)) and in Chester Beatty Codex AC. 1390 p. 1 line 2 (ναύβιον τετραγωνοειδ(ές)), where Brashear offers the meaning "trench."

24–25. In the solution to the present problem, the step halving the circumference in line 24 is a mistake since the area of a circle of circumference c is $c^2/4\pi$ which is roughly $c^2/12$, not $(c/2)^2/12$. With the correct algorithm, the volume would have been 58 1/3 volume cubits.

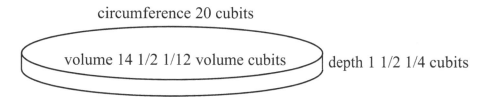

circumference 20 cubits

volume 14 1/2 1/12 volume cubits   depth 1 1/2 1/4 cubits

Fig. 41. Reconstructed diagram for problem g2. The volume ought to be 58 1/3 volume cubits.

G verso

g3: mathematical problem

Altitude and area problem: isosceles triangle.
Units: schoinia → arourai (explicitly).
Algorithm: isosceles triangle vertical algorithm B (P8B), triangle area algorithm (P6).
Diagram: same as diagram for d1.

This text is a duplicate of d1.

g4: mathematical problem

Volume problem, rectangular parallelepiped? but said to be round.
Units: cubits, fingers → cubits × fingers^2 → unnamed volumetric units and volumetric fingers.
Algorithm: parallelepiped volume algorithm (S1).
Diagram: schematic drawing suggestive of a cylinder, damaged, labelled with numerals and names of dimensions.

10. The damaged noun at the beginning is likely to have been ναύβιον, cf. g2, G recto 22. As in that problem it is said to be "circular," and the diagram appears to be of an elongated cylinder. But the fact that the given dimensions include a large "length" and a small "width" and "thickness" do not fit a circular or cylindrical shape, and the solution proceeds according to the algorithm for a rectangular parallelepiped (Fig. 42). For the metrology, see problem a3. ἴσον appears to make no sense, and perhaps is a garbling of the same word as μισ ͺων[ ] in the corresponding position at G recto 22.

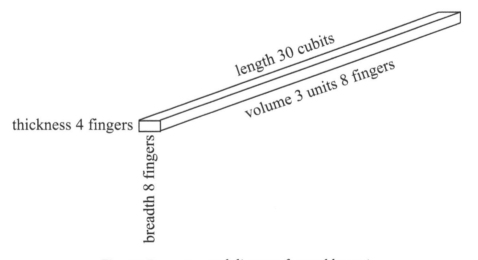

Fig. 42. Reconstructed diagram for problem g4.

M recto

m1: metrological text

This text is the most problematic of the metrological texts in *P.Math.*, involving textual and metrological puzzles that we have found intractable. It falls into five sections: (i) 1–4, severely damaged but apparently concerning units of length, (ii) 5–9, concerning units of area, (iii) 9–13, concerning units of volume, (iv) 13–14, concerning units of area, and (v) 14–19, concerning units of length.

3. For the βαρβαρικὴ σχοῖνος, equivalent to 45 stades, see the metrological list published by Heiberg as pseudo-Heron, *Geometrica*, 196.14.

4–9. Text g1, G recto 5–6 is a highly abridged version of 4–7. Outside of the present text and g1, laura ("lane"), amphodon ("street" or "block"), and dodekatikon ("relating to twelve"?) are, to our knowledge, unattested as names of units of any kind, and δωδεκατικός is in neither *LSJ* nor the *TLG*. In text g1, G recto 5–6, all three units are mentioned but not defined, as part of a list that otherwise consists entirely of units of length. Here, however, the amphodon is explicitly defined as an area unit, equivalent to a rectangle 100 cubits by 200 cubits or 20,000 area cubits in total; and since a relation is given between the dodekatikon, laura, and amphodon, all three should be area units. If so, the 10,000 cubits in 5–6 are area cubits, and the laura is identical to the urban aroura (cf. e3, E verso 20–21).

We are not sure of the function of the verb καλοῦντε (*scil.* καλοῦνται) or its parallel καλῖται (καλεῖται) at G recto 6. Syntactically, the easiest solution is to restore 4–5 as τὰ σ̄μ̄ σχοινία σχ[οῖνος, ἢ] | [δω]δεκατικὸν καλοῦντε, "240 schoinia are called a schoinos or dodekatikon," but in G recto 6 the syntax seems to call for restoring a relative clause. We confess to some discomfort in adding an otherwise unattested use of dodekatikon as an alternate name for the length unit schoinos to its likewise unparalleled use three lines later as an area unit.

9–10. The definition of the naubion as a square xylon with depth one xylon is roughly paralleled in e3, lines 13–14.

10–12. Deduction of the relation between the naubion and the volume cubit. Paralleled in e3, lines 14–15.

12–13. Relation between the volume cubit, artaba, and metretes. Paralleled in e3, lines 16–18, where the number of metretai in a volume cubit is not omitted.

13–14. Relation of urban and country aroura to bikos. The same information is in e3, lines 18–22, but this is not a textual parallel.

14–19. Length units. A schoinos comprising 4 miles (14–15) must be the conventional 30-stade schoinos; cf. pseudo-Heron, *Geometrica* ed. Heiberg 196.5–13 and 402.20. The equation of 1 mile with 7 1/2 stades implicit in these relations was itself standard (*ibid.* 194.20 and 402.17). Instead of directly expressing the mile as a number of stades, however, 15–16 state that one mile comprises 3 units called gues ("field"), each of which in turn comprise 5 stades, so that the mile is impossibly made to contain 15 stades, double the expected number. The relations between stades, surveyor's schoinia, and cubits in 16–17 are standard. Because of the mile-gues-stade doubling, the equivalents of the schoinos in the smaller units in 17–19 are twice what one would expect, namely 30 stades, 120 surveyor's schoinia, and 11,520 cubits.

M verso

m2: mathematical problem

Arithmetical problem: arithmetical sequence.
Units: artabas.
Algorithm: arithmetical sequence algorithm (N1).
Diagram: five horizontal line segments labelled with the numbers of the solution.

The statement of the problem, which is almost entirely lost, must have been to determine how to divide 300 artabas of wheat among five people so that the shares form an arithmetical sequence increasing by 8 artabas. The solution includes a final check that the sum of the sequence is correct.

N presumed recto

n1: mathematical problem

Volume problem: frustum of cone.
Units: cubits → volume cubits → naubia.
Algorithm: conic frustum volume algorithm (S3A).
Diagram: two concentric circles with a diameter drawn through them, labelled with numbers from the problem and solution.

The object (Fig. 43) may have been a circular cistern or excavation with sloping sides.

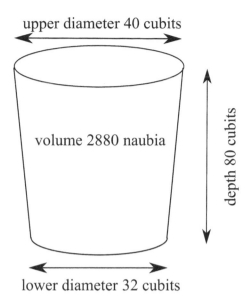

Fig. 43. Reconstructed diagram for problem n1.

n2: mathematical problem

Volume problem: triangular prism.
Units: cubits → volume cubits → artabas.
Algorithm: triangle area algorithm (P6) erroneously applied, prism volume algorithm (S2).
Diagram: right-angled triangle labelled with numbers from the problem and solution.

Aside from the mistake of taking a quarter rather than a half of the product of length and width in line 15, the solution is structurally the same as that of n3, but there are many differences in the wording.

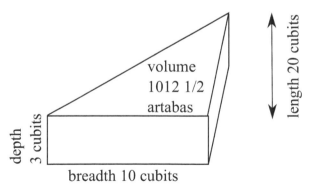

Fig. 44. Reconstructed diagram for problem n2.

N presumed verso

n3: mathematical problem

Volume problem: triangular prism.
Units: cubits → volume cubits → artabas.
Algorithm: triangle area algorithm (P6), prism volume algorithm (S2).
Diagram: right-angled triangle labelled with result of solution.

The division by two in line 3, as well as the diagram, indicate that the granary has a triangular horizontal cross section, and the "length" and "width" are the base and altitude of the triangle (Fig. 45). The factor 3 1/4 1/8 is the number of artabas in a volume cubit. Problem n2 is similar but marred by a mistake in the algorithm.

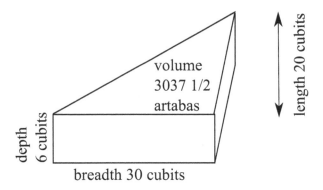

Fig. 45. Reconstructed diagram for problem n3.

n4: mathematical problem

Volume problem, compound shape.
Units: cubits → volume cubits (strictly four-dimensional cubits!) → artabas.
Algorithm: see below.
Diagram: rectangle with hatched arc along top, inscribed with result of calculation.

A similarly expressed problem, similarly solved, is o1. The vaulted granary whose capacity is to be found was probably supposed to be a chamber in the form of a rectangular parallelepiped—hence the three given orthogonal dimensions "length," "width," and "depth"—surmounted by a barrel roof.[6] If the cross-section of the roof was understood as a semicircle, its area—and hence the volume of the vault—could be calculated knowing that the given width as the semicircle's diameter (adapting algorithm P10A, multiply the width by itself, subtract a quarter of the result, and divide by two). In n4 and o1, however, we are given a separate dimension called the "vault," which we suppose refers to the height of the vault, measured from the lowest to the highest

6. See for example the vaulted granaries at Karanis, Husselman 1952.

point of the roof. Perhaps the idea was that the cross-section might be a half-oval rather than an exact semicircle (Fig. 46), so that its area could be calculated as three-quarters of the width times the vault height.

In n4, the given vault dimension is greater than either the width or the depth of the granary proper, whereas in o1 it is equal to the depth and two-thirds of the width. A roof having either of these heights would be implausibly elongated vertically. We suspect that this merely reflects the tendency in these didactic exercises to choose quantities without concern for what one was likely to encounter in the real world. (For what it is worth, n4's diagram suggests that the vault's roof is a flattish half-oval, while that of o1 defies interpretation.)

The foregoing, however, remains in the realm of speculation. Instead of following any procedure appropriate for some solid figure comprising a base portion and a roof portion, the solver absurdly multiplies all four given dimensions pairwise, the length by the width and the depth by the "vault," and then multiplies these products together. The result (in "hyper-cubits"?) is treated as if it was a volume in solid cubits, which the solver converts to artabas using the usual relation 1 solid cubit = 3 1/4 1/8 artabas.

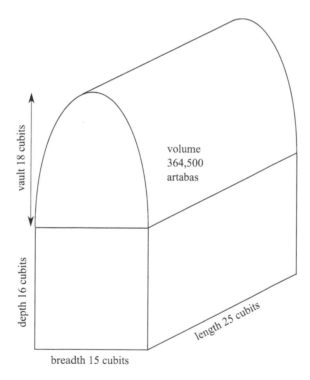

Fig. 46. Reconstructed diagram for problem n4. The drawing represents our understanding of the intended scenario, but the volume is the incorrectly computed one from the text.

O recto

o1: mathematical problem

Volume problem: compound shape.
Units: cubits → volume cubits (strictly four-dimensional cubits!) → artabas.
Algorithm: see below.
Diagram: a strange drawing, with a notional quadrilateral having convex right and left sides and concave lower side, and above the upper side a zigzagging line. A few numerical labels.

For a similar problem see n4. The statement and solution are terser here, and instead of multiplying the four dimensions pairwise (length by width, vault height by depth), a single product is built up by multiplying the length (miscalled the "width" at the beginning of line 4) by the depth, width, and vault height in that order.

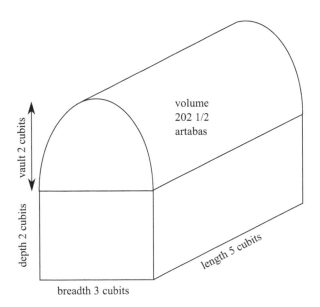

Fig. 47. Reconstructed diagram for problem o1. The drawing represents our understanding of the intended scenario, but the volume is the incorrectly computed one from the text.

o2: mathematical problem

Linear dimension from area problem: quadrilateral.
Units: schoinia, arourai → schoinia.
Algorithm: inverse quadrilateral area algorithm A (P4i).
Diagram: horizontal line labelled with the lengths of the four sides. For a similar diagram see problem o5.

In Chester Beatty Codex AC. 1390, p. 3, 1–12, we are given the area and the length of *one* side of a quadrilateral, and the problem is to find the lengths of the other three sides assuming that the quadrilateral area algorithm (P4) applies exactly. That problem is indeterminate, and the solution offered (see inverse quadrilateral area algorithm B) is one of an infinity of possibilities. In the present problem, we are more reasonably given the area and *three* sides, making the problem determinate. The choice of givens, however, is disastrous (Fig. 48). It is easy to see that the fourth side of any quadrilateral having three sides of length 4 must be less than 12, and that such a quadrilateral must have a smaller area than a trapezoid with parallel sides 4 and 12 and altitude 4, i.e., less than 32. Applying inverse quadrilateral area algorithm A to the given data leads to the absurd result that the fourth side is the sum of the other three sides, so that the "quadrilateral" is a straight line of area zero. (Presumably the diagram is not meant to illustrate this result but is just a schematic layout of the data, as in problem o5.)

It is also surprising that the three given sides have been chosen to have equal length, so that it is not clear to which sides the two 4s occurring in line 11 of the solution pertain. In fact the first 4 represents the average of the two given sides that happen to be opposite; since they are equal, the step of calculating this average was omitted. The second 4 represents the third given side, which is opposite the unknown side.

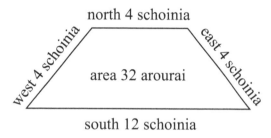

Fig. 48. Reconstructed diagram for problem o2. The diagram represents the quadrilateral as we suppose the originator of the problem imagined it.

o3: mathematical problem

Area problem: uncertain shape.
Units: cubits → area cubits → bikoi
Algorithm: see below.
Diagram: none.

We take ψιλός to be equivalent to ψιλὸς τόπος, "vacant lot." Three dimensions are provided: a length, a width, and a third dimension which may or may not have been named. In the solution, the three dimensions are multiplied together as if one was calculating the volume of a parallelepiped, but the product is treated as area cubits, and converted to bikoi, an area measure. Perhaps the figure was intended to be a trapezoid.

o4: mathematical problem

Geometrical problem: sides of a triangle.
Units: none.
Algorithm: ?
Diagram: none.

This problem and its solution—if these are not merely disconnected calculations—are incomprehensible.

O verso

o5: mathematical problem

Area problem: quadrilateral.
Units: schoinia → arourai.
Algorithm: quadrilateral area algorithm (P4), modified.
Diagram: horizontal line labelled with the lengths of the four sides. For a similar diagram see problem o2.

6. The sum of 5 1/2 1/4 1/8 and 4 1/2 should have been 10 1/4 1/8. The figure 5 1/16 given in line 7 for half this sum implies that the 1/4 was accidentally dropped, leaving 10 1/8. Instead of 1/8, however, the sloppily written fraction at the end of line 6 is either 1/2 or (less likely) 1/6.

7–10. Before multiplying together the averages of opposite sides, the solver multiplies each by 8. Their product is then divided by 64, cancelling out the previous operations. The point of these mathematically superfluous steps was probably to simplify the multiplication of the fractions. It is not clear whether ἀπόδιξεις ("demonstration"?) in line 7 signals this arithmetical procedure.

The solution from this point is marred by more serious errors. Apparently the original attempt at solution went as follows:

7 1/2 1/4 1/6 × 8 = 62 1/2 (correct)
5 1/16 (the solver's miscopying of 5 1/8 1/16) × 8 = 40 1/2 (correct)
62 1/2 × 40 1/2 = 2531 1/4 (correct)
2531 1/4 ÷ 64 = 39 1/2 1/32 1/64 1/256 (correct)

In other words, the solver accidentally dropped part of the fractions for the average of the east and west sides, and finished the calculations using this false number but making no further errors. Thereafter, whoever copied the text added a spurious 2, making 40 1/2 into 42 1/2 in both its occurrences in line 9, but this mistake did not affect the subsequent calculations.

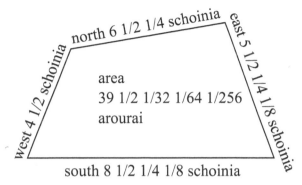

Fig. 49. Reconstructed diagram for problem o5. The area is the erroneous value obtained in the text.

o6: mathematical problem

Arithmetical problem: wages.
Units: talents.
Algorithm: proportions.
No diagram.

Despite the fact that most of the text of this problem survives, a full reconstruction of it eludes us, and for want of parallel texts, our interpretation and restorations must be tentative. The scenario involves workers—brush-cutting is mentioned in line 12, and river-workers in 15—numbers of work days, and pay in talents. The arithmetical operations on the given quantities apparently begin in line 16, and are as follows:

(i) 6×15 (days) = 90
(ii) 8×15 (days) = 120
(iii) 120×120 talents = 14,400 (talents)
(iv) $14,400 \div 90 = 160$ talents

160 talents, as an amount that "they will receive," is explicitly indicated as the solution of the problem. The nature of the problem, however, is not so easy to discern. First, what do the numbers 6 and 8 in steps i and ii represent? If they are daily wages of two distinct categories of individual worker, then the products 90 and 120 obtained in these steps would be the total pay per worker over the 15 days of labor, presumably in talents though the unit is not mentioned in these steps. This interpretation gives rise to several objections. If the figure 120 talents is arrived at as part of the solution in step ii (line 16), why is it already stated in 13, apparently as one of the givens? What meaning can a multiplication of 120 talents by 120 talents in step iii have? And if 90 and 120 are total payments per worker for two kinds of worker, to whom does the 160 talents obtained as the solution apply?

Alternatively, 6 and 8 could represent numbers of workers. In this case, the 120 talents given in line 13 and the 160 obtained as the solution are daily wages of individual workers, the

90 and 120 obtained in steps i and ii are numbers of worker-days respectively for a group of 6 and of 8 workers, and the 14,400 talents is the total outlay for the group of 8, thereafter assumed to be also the total outlay for the group of 6. The circumstance that 120 is both the number of work-days performed by the group of 8 workers and the daily wage that they receive would be a coincidence. (This makes sense of the way the multiplication in 18 is expressed, since a squaring operation on a single quantity would normally be written as ἐφ᾽ ἑαυτά, "by itself.") We could paraphrase the problem as: "If 8 workers are paid 120 talents per day for working 15 days, how much would 6 workers be paid for working 15 days if the total sum paid remains the same?"

The computation of days of work as a kind of abstract measure is well attested in the papyri. A good example occurs in *BGU* 1.14, which is cited below in the note to l. 15. To quote from the translation in Johnson 1936, 216–219, "35 river men on full time under Sarapammon at 9 dr. Employed as follows: 20th, 4; 21st, 7; 22nd, 9; 23rd, 10; 24th, 5." So 35 river men are in fact 35 river men days of work, not 35 actual individual men. This sort of abstraction is also common in the Heroninos archive, which is contemporary with this papyrus and is discussed in Rathbone 1991: 391–392. The wage of 120 talents per day may strike the reader as an unreasonably high day, but in the period to which we assign the composition of this codex, it is not. In *PSI* 4.287, an apprenticeship contract dated to 377, an apprentice *tarsikarios* (a kind of weaver) is to be paid 200 talents per day. Given some degree of price rise since the mid-360s, 120 talents here seems reasonable. (See Bagnall 1985, 46 on the difficulty of tracking the degree of price rises between the 360s and the 380s, with specific reference to this papyrus.)

Although the foregoing looks like a plausible scenario that would account for the arithmetical operations, making sense of line 17 remains a serious difficulty. If πολῖτε represents πωλεῖται and is not just a blunder of the writer, what is being sold for 120 talents in the context of a problem that seems otherwise to be about wages?

12. The initial word of the line might be [ἐργάτη]ς, "a worker," or perhaps [ποταμίτη]ς, "river worker," although that may be too long for the available space.

14. It is not immediately clear whether the participial phrase belongs with the preceding phrase or with what follows. But we find it difficult to reconstruct a plausible sequence of clauses that makes it part of what follows, and we have reconstructed it as going with what precedes.

15. The sudden appearance of ποταμῖται here, if it did not already occur in l. 12, is surprising, and it might suggest that we are dealing with a second group of workers, after the "wood-cutters" suggested by the participle in l. 12. But matters are more complicated. The verb ξυλοτομέω appears only once in the papyri, to our knowledge, in *P.Ross.Georg.* 2.19.31, and is listed by LSJ only in the *Supplement*, from this papyrus. Like the noun ξυλοτομία, which is more common in the papyri, it regularly refers to pruning vines ("Rebschnitt": Schnebel 1925, 262–267); LSJ's definition ("woodcutting") ignores this specific usage.

But on a more general reading, the term is far from irrelevant to river workers. One of the things *potamitai* do is to use brush that has been cut to reinforce dikes or the walls of bodies of water. *BGU* 1.14, a Memphite account of 255 CE, provides some good examples. The verb for "reinforcing" there is παρυλίζοντες, which LSJ renders "make dykes with brushwood." All in all, it does not seem at all implausible that *potamitai* would cut brush. As they do a total of 95 days of work in *BGU* 1.14, col. iii, the kind of numbers we are dealing with in this problem (120 or 90 man-days of work) are also not at all unreasonable.

17. ἀνὰ λόγον (or ἀνάλογον) should indicate a relation involving proportionality. In the present context, if this is a calculation of daily wages for different numbers of workers where the total amount of money is constant, the wages are in inverse proportion to the numbers of workers. Beyond that, the syntax of this line is obscure, and (as observed above) πολῖτε = πωλεῖται, "is/ are sold" seems inappropriate (and conceivably, as suggested by one of the reviewers of this edition, a miscopying of ποταμῖται, though this may seem an appeal to *obscurum per obscurius*). Perhaps underlying what we have is a thought along the lines, "some amount is the wage for the 90 (worker-days) in (inverse) proportion as 120 talents is the wage for the 120 (worker-days)."

18. η̄ (8) is an isolated error for ϙ̄ (90), as the result of the division in the next line makes clear; the mistake could have arisen from a misreading of a carelessly written qoppa, reinforced perhaps by the legitimate appearance of 8 previously as a number of workers.

H →

h1: metrological text

As far as line 16, this text is a rigorously systematic verbal tabulation of a system of units of length of decreasing magnitude, giving the value of each unit in terms of all the smaller units in descending order. The text can be extropolated backwards to the beginning of the section on the schoinion two lines before the first preserved line, about halfway across the line. It is unlikely that the list began with a still larger unit, since that would require at least three additional lines, so we need only suppose a brief introductory phrase like that at the beginning of g1. Lines 1–2 (and the reconstructed preceding lines) are paralleled in *P.Lond* 5.1718.79–80, with some scrambling of the order of units. Lines 2–3 and 11–14 are paralleled, in a tabular layout, in *P.Ryl* 2.64.

6. Perhaps the copyist miswrote [ψιμα]θάς; without doubt a spelling mistake must be supposed.

11. The number of palms in a cubit should be 6. The trace at the end of line 6 looks like the top of a delta. One might conjecture δ[ε̣ ϛ̄], but in that case the overstroke marking the numeral is extended too far to the left.

16. The new list beginning here is evidently supposed to be composed of volumetric units, since this is the only way to make sense of the inclusion of the naubion and its equivalence to 27 (volume) cubits. The other units in the list have the same proportion in volume to the volume cubit as the length units bearing the same names have to the linear cubit. It is worth remarking that the volumetric spithame according to this system has the same magnitude as the unnamed unit in problems a3, c1, and g4 while the volumetric lichas has the magnitude of the unnamed unit in problem b5; however, the volumetric fingers in those problems are 24ths of their unnamed units, not of the volume cubit.

h2: partition into unit-fractions

See introduction, section 10 for problems of this type, in the present instance too badly preserved for separate discussion.

H ↓

h3: model contract

H ↓ is devoted in its entirety to an undertaking to lease arable land.[7] The landowner appears in lines 1–2, the prospective lessee in lines 2–3. The prospective lessor resides in Oxyrhynchos, the lessee in the hamlet (epoikion) of Hiereon ("of the priests"), in the 6th pagus. Abrasion of the upper right part of the sheet has removed much of the text describing the parties and the location of the land at stake, but most of the rest is readable or restorable. The formula used here, an undertaking with ἑκουσίως ἐπιδέχομαι, is typical of the Oxyrhynchite nome: see Herrmann 1958, 41 with n. 4.

1 The nomen, or first name, Flavius, appears in the fourth century with the names of persons holding imperial status of some kind, usually military or civil office. See Keenan 1973–1974 and 1983. As Constantine's family name, it is found in papyri once he took control of Egypt at the end of 324 and retains its position to the end of Roman rule. It thus provides only a *terminus post quem*. The lessor should by virtue of the name have been a person of some consequence, but we have not managed to read enough of the right side of line 1 to determine if ὑπα- is part of a patronymic or rather refers to some status. It is hardly possible to suppose that the (presumably fictitious) Diogenes was consul. In any event we have not found any individual named Flavius Diogenes in other texts.

3 The ἐποίκιον Ἱερέων is known from documents ranging from the late third to the sixth/seventh century; see Pruneti 1981, 68; Benaissa 2012, 111–112,[8] hazarding the suggestion of

---

7. Several leases appear in the Panopolite codex BL Add. MS 33369.

8. Benaissa adds *P.Oxy.* 7.1072.14 to Pruneti's list, citing *BL* 10.140, where J. Gascou's observation (*P.Sorb.* 2, p. 66 n. 108) that in the Oxyrhynchite text in question one should read the placename Ἱερέων rather than the

location in or near the 3rd pagus. Its correct assignment to the 6th pagus was not previously known. This pagus was located in the middle of the nome, overlapping with the old Middle Toparchy, a little north of the latitude of Oxyrhynchos itself: see the map in Rowlandson 1996, xiv, with discussion on 13. What follows the naming of the pagus here is read with reserve. One might anticipate τοῦ (αὐτοῦ) Ὀξυρυγχίτου νομοῦ, but the traces do not support that, and space is insufficient. The proposed reading is slightly elliptical. For τω as an error for τοῦ cf. l. 6. Real topographical names also appear in the Panopolite codex.

4–5 In principle, a year 10 could refer either to 342/3 (deceased Constantine I 37, Constantius II 19, Constans 10) or to 364/5 (41=10, Oxyrhynchite era years by the deceased Constantius II and Julian). For the first, see Bagnall and Worp 2004, 56; for the second, 59. Apart from the consideration that era years appear from the 360s on much more regularly than regnal years do in the period after Constantine's death, there is a strong presumption in favor of the 360s as the "dramatic date" of the documents in this codex, based on the amounts of money in both A recto and I ↓, which would not occur in documents of this type before 350. A date referring to 364/5 thus seems highly likely.

6 On chorion see Bagnall 1999. In ἐδαφ, the last two letters are abraded; there may have been a correction, as there seems to be one too many diagonal strokes. Presumably the genitive plural ἐδαφῶν is needed, but we cannot confirm ων following, and the writer's shaky command of case may have led him to write some case other than the genitive. We cannot suggest an interpretation for the following θρ.

10 πενταμναῖος has occurred in the papyri otherwise only (as the editor there notes) in P.Rain. Cent. 86.12,24 (a reedition with a new fragment of SPP 20.103), where the neuter plural is spelled πενταμνιεα. There it refers to tow, from the flax plant, as here, and also modifies δεσμίδια, bundles. LSJ records the word only with a meaning of "five months."

12 ἐφ' ὑημσίας (l. ἡμισείας): see O.Kellis 145, where it refers to the basis of a shared purchase of a cow, or κοινῇ ἐφ' ἡμισείας in P.Cair.Masp. 2.67159. In P.Oxy. 6.913, the phrase describes, as here, the division of the produce of the fields between lessor and lessee. ἐξ ἡμισείας is used similarly. For τῶ περεὶ (περὶ) κινῶ (κοινῷ), on the other hand, we have not found a parallel, and it is not at all clear what stood in the lacuna between this phrase and the start of the next clause in line 13 (ἀκίνδυνος …).

13 It does not appear that καὶ ἐπερωτηθεὶς ὡμολόγησα was written at the end here, as one would expect.

---

common noun "priests" is recorded. But on the next page Benaissa refers to the presence of a Christian priest, without making it clear that he is (presumably) alluding to the Apa Martyrios, presbyteros, to whom the letter is addressed. There is no reason to suppose that he was in epoikion Hiereon.

I →

i1: metrological text

In contrast to h1, this text is structured hierarchically: short lists of relations between pairs of units of the same class, progressively descending in magnitude, lead quasi-syllogistically to a concluding relation between the largest unit and the smallest unit or units. Each such list is independent of the lists that precede and follow, so the large scale order is arbitrary. There are several parallels to h1, of which the closest and most extensive is *P.Oxy* 1.9+49.3456; others are *P.Oxy* 49.3457–3460. These texts are composed of the same kinds of lists as i1, with the lists often verbally identical or nearly so but with variation in the order of the lists.

[0]–2. Length units: cubits, palms, and fingers. Parallels: *P.Oxy* 1.9, 11–13; *P.Oxy* 49.3457, 8–10; *P.Oxy* 49.3458, 13–16; *P.Oxy* 49.3459, 6–9. The lost line preceding line 1 likely began with an introductory phrase like that of g1.

2–4. Liquid capacity units: metretai, choeis, and kotylai. Parallels: *P.Oxy* 1.9, 13–14; *P.Oxy* 49.3457, 3–5; *P.Oxy* 49.3458, 17–19.

4–7. Weight units: mnaeia, tetartai, thermoi, carats. Parallels: *P.Oxy* 49.3456, 15–17; *P.Oxy* 49.3457, 10–11; *P.Oxy* 49.3460.

7–10. Weight units: litrai, unciae, semiunciae, grammata. Parallel: *P.Oxy* 49.3456, 19–21. The restored text at the end of line 9, required by the context and paralleled in *P.Oxy* 49.3456, appears to be too long for the available space; perhaps a word was accidentally omitted.

10–14. Weight units: talents, minas, staters, drachmas, tetrobols. Parallel: *P.Oxy* 1.9, 4–7; *P.Oxy* 49.3458, 2–7. In line 10, the superfluous εξ followed by $\overline{\xi}$ corresponds to [[ε]]$\overline{\xi}$ in *P.Oxy* 1.9, 5, an indication of a not-too-distant common text. In *P.Oxy* 1.9, an equation of 1 drachma with 7 obols appears as part of an earlier grouping of drachmas, obols, and chalkoi in lines 2–4 (and this was likely true of *P.Oxy* 49.3458 too), but there is nothing to correspond to the equation here of 1 stater with 7 tetrobols. The circumstance that the text gives the equivalent of a stater in tetrobols but that of a talent in obols might indicate a lacuna, though the omitted equivalence of 1 tetrobol with 4 obols is tautological.

i2: mathematical problem

Volume problem: rectangular parallelepiped.
Units: cubits → volume cubits → metretai and xestai.

Algorithm: parallelepiped volume algorithm (S1).
Diagram: none.

The metrological conversions depend on the equivalence of 1 volume cubit to 3 metretai, and of 72 xestai to 1 metretes. The relation of metretai to volume cubits is given in two of the metrological sections, e3 (E verso 17–18) and m1 (M recto 13), in the latter passage with the numeral "3" erroneously omitted. Xestai are not mentioned elsewhere in the codex. The solution to this problem appears to be first to convert the capacity of the vat from volume cubits into metretai, then to determine the number of xestai in one volume cubit, and lastly to use this factor to convert the number of volume cubits directly into xestai, a rather inefficient procedure.

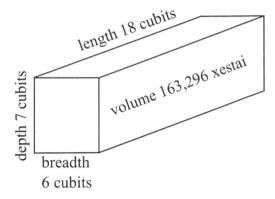

Fig. 50. Reconstructed diagram for problem i2.

I ↓

i3: model contract

Like A recto.1–12, this passage contains the remains of a loan of money. Here again, the information about the parties is largely lost at the beginning, although we have the patronymic of one party, Hatres, in line 3, and the information that he comes from a village, of which the name is lost. It would be unsafe to assign οἰκῶν to the borrower on the basis of the nominative case; both second position and a village residence suggest that the person whose description probably began toward the end of line 2 and ended in line 4 was the borrower. As with the other loan, no consular date was provided at the start, nor subscription at the end.

The text is badly damaged, especially at right, and it is not possible to restore it to anything like its original full extent. But it is clear from line 5 that the loan belongs to the category of those made in money but to be repaid in a commodity, in this case wool. Such contracts, although from the borrower's point of view entered into primarily to obtain working capital or living expenses on credit, are hybrids containing also elements of sale. They are for this reason often called "sales on delivery" (German Lieferungskäufe). Such documents are discussed in detail by A. Jördens in *P.Heid.* V, pp. 296–341. Examples appear also in the Panopolite codex and in *P.Yale* 4.186 and 188. Wool is an unusual object of such a transaction, with Jördens listing only one example, *P.Sakaon* 95 (Theadelphia, 301), where the amount is specified as 37 pounds of clean, spun wool. The price is specified, as is the duration of the contract (nine months, with a due date in Pharmouthi, or two months later than in our contract). The formula of this document is rather different from ours.

4 Most likely we should restore [αὐτο]ῦ, which suits the space and would mean that the name of the nome, presumably the Oxyrhynchite, had been mentioned in line 2. But it is also possible that as is apparently the case in H ↓ 3, the lender's identification instead referred to the city of Oxyrhynchos. The beginning of line 5 suggests that we may have had in line 4 the beginning of a variant version of the "need" formula typically found in loans, εἰς τὴν ἰδίαν μου καὶ ἀναγκαίαν χρείαν.

6–7 The amount of money received as the value of the wool is mostly lost. If the letters ψν that we have read in line 7 are the restatement of the amount in numbers, we should probably restore δηναρί[ων μυριάδας ἑπτακοσίας πεντήκοντα], | [γί(νονται) (δηναρίων) (μυριάδες)] ψν, i.e., 750 myriads of denarii, or 7,500,000 denarii, or 5,000 talents. If, as seems likely, the date is around 364/5, this would amount to something like the value of 5 artabas of wheat. Given the loss or omission of the amount of wool at stake, it is impossible to evaluate the plausibility of this amount. It is unclear what the letters at the start of l. 6 refer to; we cannot find a parallel for a compatible term modifying ἀργυρίων (which is itself uncommon). The neuter form of ἅπερ in l. 7 should not give us undue concern, as lack of agreement in such clauses is common.

8 The remains of the last letter of the line are compatible with mu. In the Oxyrhynchite nome, a regnal or era year beginning with mu (40) would have been possible from 345/6 to 354/5 (40–49 to the deceased Constantine and regnal years 22–31 of Constantius II with successively Constans and Gallus providing a third number) and again with the Oxyrhynchite era in its long-term form, from 363/4 to 372/3 (posthumous years of Constantius II and Julian). In H ↓.5 we find the end of a date, to year 10 (see note ad loc. for references). That would not be possible as the final year in the sequence from the first series of dates and can refer only to year 41=10, or 364/5. It is possible that the same year was written here, but it cannot be excluded that another year in the same general period was used.

i4: partition into unit-fractions

See introduction, section 10 for problems of this type. The present instance is too damaged for separate discussion.

i5: mathematical problem

Too little remains of this problem (or problems?) to recover the sense.

Appendix A. Synopsis of Problem Texts in Mathematical Papyri

In the following tabular summary of the problem texts in Greek mathematical papyri, the columns are as follows: (i) the ordinal number of the problem within the text, (ii) the relevant lines of the edition, (iii) the designation of the object that the problem concerns, (iv) a short description of the problem, for geometrical problems including the kind of shape, the kind of quantity sought, and the units of the givens and result, and (v) the algorithm employed (from introduction, section 10).

BM Add. MS 41203A

1	r1–6	φρέαρ στρογήλουν round well	frustum, volume, cubits? → bricks	conic frustum surface algorithm (S4)
2	r7–11	ἐξέδρας πεπλακομένον paved platform	rectangle, area, unspecified length units → slabs	
3	r12–16	βουνὸς σίτου στρογήλον round heap of grain	frustum, volume, cubits → artabas	conic frustum volume algorithm A (S3A)
4	r16–19	δόξα arches	quadrilateral prism, volume, cubits? → bricks?	

Chester Beatty Codex AC. 1390

1	1.2–14	ναύβιον τετραγωνοειδές rectangular trench	quadrilateral prism, volume, cubits → naubia	quadrilateral area algorithm (P4), prism volume algorithm (S2)
2	1.15–23	θησαυρὸς στρογγυλοῦν round granary	frustum of cone, volume, cubits → artabas	conic frustum volume algorithm B (S3B)
3	1.24–27		subtraction of unit-fractions from 1	
4	2.1–6	λίθος stone	parallelepiped, volume, cubits → volume cubits	parallelepiped volume algorithm (S1)
5	2.7–13	plots of land	proportional shares of removed portion	
6	2.14–27	σφραγίς plot of land	quadrilateral (actually rectangle), area, schoinia and cubits → arouras and hammata	quadrilateral area algorithm (P4)
7	3.1–12	σφραγίς plot of land	quadrilateral, linear dimension, reeds and arouras → reeds	inverse quadrilateral area algorithm B (P4iB)
8	3.13–18	σφραγίς plot of land	quadrilateral, area, reeds → arouras	quadrilateral area algorithm (P4)
9	3.19–23		Prorating of amounts in talents	
10	3.24–4.9		Prorating of amounts in myriads of denarii	

MPER NS 1.1

1	1.23–2.4		metrology	
2	2.4–7	κύβος cube	cube, volume, feet → volume feet	parallelepiped volume algorithm (S1)
3	2.8–13	λίθος stone	cube, volume, feet → volume feet	parallelepiped volume algorithm (S1)
4	2.13–3.6		metrology	

5	3.7–12		cube, volume, fingers → volume fingers	parallelepiped volume algorithm (S1)
6	3.12–15		cube, volume, feet? → volume palms	parallelepiped volume algorithm (S1)
7	3.15–4.3		metrology	
8	4.3–8	κύβος cube	cube, volume, feet → volume feet	parallelepiped volume algorithm (S1)
9	4.9–11	ἄσκωμα tube?	parallelepiped, volume, feet → ?	
10	5.1–5	πυραμίς pyramid	equilateral triangular pyramid, volume, feet → volume feet	pyramid volume algorithm A (S5A)
11	5.6–18	πυραμὶς τρίγωνος ἰσόπλευρος κολούρα truncated equilateral triangular pyramid	equilateral triangular pyramidal frustum, volume, feet → volume feet	pyramidal frustum volume algorithm A (S6S)
12	5.19–6.3	στήλη block	equilateral triangular pyramid, volume, units unclear	pyramid volume algorithm A (S5A)
13	6.3–11	πυραμὶς τρίγωνος ἡμιτελής half-finished triangular pyramid	equilateral triangular pyramidal frustum, volume, feet → volume feet	pyramidal frustum volume algorithm A (S6A)
14	7.1–8	πυραμὶς ἀμβλυγώνιος obtuse-angled pyramid	equilateral triangular pyramidal frustum, volume, feet → volume feet	pyramidal frustum volume algorithm A (S6A)
15	7.9–12	πυραμὶς τρίγωνος triangular pyramid	triangular pyramid, volume, feet → volume feet	triangle area algorithm (P6), pyramid volume algorithm C (S6C)
16	7.12–16	λίθος, τρίγωνον τῇ ἐπιφανείᾳ stone, triangle in surface	triangular prism, volume, feet → volume feet	triangle area algorithm (P6), prism volume algorithm (S2)
17	7.16–18		too damaged to interpret	

18	8.1–12	πυραμὶς τετράγωνος ὀξεῖα acute quadrilateral pyramid	square pyramid, volume, feet → volume feet	pyramid volume algorithm B (S5B)
19	8.12–19	πυραμὶς τετράγωνος κολλούρα truncated quadrilateral pyramid	square pyramidal frustum, volume, feet → volume feet	pyramidal frustum volume algorithm B (S6B)
20	9a.1–2	κύλινδρος cylinder	cylinder, volume, feet → volume feet	circle area algorithm (P10A), prism volume algorithm (S2)
21	9a.2–5	κύκλος circle (sic)	cylinder, volume, feet → volume feet	circle area algorithm (P10A), prism volume algorithm (S2)
22	9a.5–6	ἄλλος another (scil. cylinder)	cylinder, volume, feet → volume feet	circle area algorithm (P10A), prism volume algorithm (S2)
23	9a.6–7	κύλινδρος cylinder	cylinder, volume, feet → volume feet	circle area algorithm (P10A), prism volume algorithm (S2)
24	9b.1– 10.7	κῶνος ἡμιτελής half-finished cone	conic frustum, volume, feet → volume feet	conic frustum volume algorithm D (S3D)
25	10.7–16	κῶνος ἡμιτελής, λίθος half-finished cone, stone	conic frustum, volume, feet → volume feet	conic frustum volume algorithm D (S3D)
26	11.1–10	πυραμίς, πυλών pyramid, tower	frustum of equilateral triangular pyramid, volume, feet → volume feet	pyramidal frustum volume algorithm A (S6A)
27	11.10– 18		equilateral triangular pyramid, volume, feet? → volume feet?	pyramid volume algorithm A (S5A)
28	12.1–8	τετράγωνος ἡμιτελής half-finished quadrilateral (scil. pyramid)	frustum of square pyramid, volume, feet? → volume feet?	pyramidal frustum volume algorithm B (S6B)
29	12.8–13	τετράγωνος quadrilateral (scil. pyramid)	square pyramid, volume, feet? → volume feet?	pyramid volume algorithm B (S5B)

30	13.1–4	τετράγωνος quadrilateral (*scil.* pyramid)	rectangular pyramid, volume, feet → volume feet	rectangle area algorithm (P2), pyramid volume algorithm C (S5C)
31	13.5–12		frustum of pyramid?	
32	14.1–6		pyramid?	
33	15.1–3		prism with trapezoidal cross-section, volume, feet → volume feet	
34	15.4–7	στῦλος pillar	cylinder, volume, feet → volume feet	
35	15.7–12		cone, volume, feet → volume feet	
36	15.12–15		parallelepiped	
37	16.1–2		cylinder	
38	16.3–7	βουκέφαλον ox-head (?)	prism with isosceles trapezoidal cross-section	

MPER NS 15.178

1	1.1–10	μηνίσκος crescent	arc-shaped figure, area, schoinia → arouras	similar to quadrilateral area algorithm (P4)
2	2.1–7	κύκλος circle	circle, area, schoinia → arouras	circle area algorithm B (P10B)
3	3.1–6	κύκλος circle	circle, area, schoinia → arouras	circle area algorithm A (P10A)
4	4.1–9	κύκλος circle	circle, area, schoinia → arouras	circle area algorithm C (P10C)
5	5.1–9	ἡμικύκλιον semicircle	semicircle, area, schoinia → arouras	semicircle area algorithm D (P10D)
6	6.1–9		unit-conversion?, dromos-artabas, cf. *PLond.* 2.265	

MPER NS 15.173+

1	1–6	κύκλος circle	circle, area, schoinia → arouras	circle area algorithm B (P10B, but text erroneously has division by 14, not 12)

P.Bagnall 35 (P.Cornell inv. 69)

1	1.1–16	lost	scalene trapezoid, area, schoinia → arouras	trapezoid area algorithm C (P3C)
2	1.17–27	lost	possibly continuation of preceding problem, area, schoinia → arouras	
3	2.1–21	τραπεζοειδής trapezoid	obtuse trapezoid, area, schoinia → arouras	trapezoid area algorithm D (P3D)
4	2.22–40	σφραγίς plot of land	quadrilateral, area, presumably schoinia → arouras	inverse diagonal rule (P1i), right-angled triangle area algorithm (P7)

P.Berl. 11529v

1	1.1–10	[παραλληλόγραμμον] [parallelogram]	rectangle, linear dimension and area, schoinia → schoinia and arouras	diagonal rule (P1), rectangle area algorithm (P2)
2	1.11–19	[τρίγωνον ὀρθογώνι]ον [right-angled triangle]	right-angled triangle, area, schoinia → schoinia and arouras	diagonal rule (P1), right triangle area algorithm (P7)
3	2.21–32	τρίγωνον ἰσόπλευρον equilateral (*sic*, for isosceles) triangle	isosceles triangle, linear dimension and area, schoinia → schoinia and arouras	isosceles triangle vertical algorithm (P8B), triangle area algorithm (P6)

| 4 | 2.33–35 | λίθος
stone | parallelepiped, volume, cubits, feet → volumetric feet | parallelepiped volume algorithm (S1) |
| 5 | 2.36–41 | τρίπους ἡρπασμένος
seized (?) tripod | unclear shape, volume?, cubits and fingers → volumetric (?) cubits and fingers | unidentified algorithm |

P.Cair. cat. 10758 (Achmim mathematical papyrus)

1	3v1.1–7	λάκκος στονγύλουν round cistern	frustum, volume, cubits → naubia	conic frustum volume algorithm B (S3B, with erroneous division by 36 likely through confusion with S3C)
2	3v1.8–12	θησαυρὸς τετράγωνος rectangular granary	parallelepiped with square horizontal section, volume, cubits → artabas	parallelepiped volume algorithm (S1)
3	3v1.13–20		proportional shares	
4	3v1.21–26		proportional shares	
5	3v1.27–30	διόρυξ τετράγωνος square trench	supposedly parallelepiped, volume, cubits → naubia	bungled algorithm
6	3v2.1–5		subtraction of unit-fractions from unit-fractions	
7	3v2.6–10		subtraction of unit-fractions from 2/3	
8	3v2.11–16		subtraction of unit-fractions from 2/3	
9	3v2.17–21		subtraction of unit-fractions from 2/3	
10	4r1.1–9		fractional shares of a house sale	

11	4r1.10–18		shares of a tax	
12	4r1.19–27		subtraction of unit-fractions from 2/3	
13	4r1.28–37		contents of a granary before deductions	
14	4r2.1–5		subtraction of fractions from 1	
15	4r2.6–9		subtraction of fractions from 1	
16	4r2.10–17		expression of a quotient as unit-fractions	
17	4r2.18–29		contents of a granary before deductions	
18	4r2.30–42		expression of a quotient as unit-fractions	
19	4v1.1–12		expression of a quotient as n unit-fractions	
20	4v1.13–20		expression of a quotient as n unit-fractions	
21	4v1.21–27		expression of a quotient as unit-fractions	
22	4v1.28–33		expression of a quotient as unit-fractions	
23	4v2.1–6		fraction of unit-fractions	
24	4v2.7–14		subtraction of unit-fraction from unit-fractions	
25	4v2.15–27		multiplication of unit-fractions, followed by subtraction of unit-fractions	
26	4v2.28–31		proportions, price of wheat	
27	4v2.32–35		proportions, price of wheat	

28	4v2.36–41		proportions, price of wheat	
29	5r1.1–5		subtraction of unit-fractions from unit-fractions	
30	5r1.6–10		subtraction of unit-fractions from unit-fractions	
31	5r1.11–15		subtraction of unit-fractions from unit-fractions	
32	5r1.16–18		subtraction of unit-fractions from 1	
33	5r1.19–21		proportions, price of wheat	
34	5r1.22–24		proportions, price of wheat	
35	5r1.25–27		proportions, price of wheat	
36	5r2.1–5		proportions, price of wheat	
37	5r2.6–11		proportions, price of wheat	
38	5r2.12–16		fraction of unit-fractions	
39	5r2.17–27		fraction of unit-fractions	
40	5v1.1–11		expression of a quotient as unit-fractions	
41	5v1.12–13		proportions	
42	5v1.14–15		proportions	
43	5v1.16–17		expression of a quotient as unit-fractions	
44	5v1.18–19		proportions, price of wheat	
45	5v1.20–21		proportions, price of wheat	

46	5v1.22–23		proportions, price of wheat	
47	5v2.1–14		three θησαυροί, proportional deductions	
48	5v2.15–25		three θησαυροί, proportional deductions	
49	6r1.1–10		three θησαυροί, proportional deductions	
50	6r2.1–15		expression of quotient as n unit-fractions	

P.Chic. 3 (P.Ayer)

1	1.1–8		poorly preserved, unedited	
2	1.9–13		poorly preserved, unedited	
3	2.1–2	[τραπέζηον ἰσοσκελές] [isosceles trapezoid]	isosceles trapezoid, area, schoinia → arouras	presumably trapezoid area algorithm B (P3B)
4	2.3–15	τραπέζηον σκαληνόν scalene trapezoid	scalene trapezoid, area, schoinia → arouras	trapezoid area algorithm C (P3C)
5	3.1–15	παραλληλόγραμμον parallelogram (actually obtuse trapezoid)	obtuse trapezoid, area, schoinia → arouras	trapezoid area algorithm D (P3D)
6	3.16–20	ῥόμβος rhombus	rhombus, area, schoinia → arouras	inverse diagonal rule (P1i), right-angled triangle area algorithm (P7)

P.Col. inv. 157a

1	A.1–11		quadrilateral, area, schoinia → arouras	quadrilateral area algorithm (P4)
2	A.12–17	σφραγίς plot of land	quadrilateral, area, schoinia → arouras	
3	B.1–17	σφραγίς plot of land	rectangle, area, hammata → arouras	quadrilateral area algorithm (P4)

P.Gen. 3.124

1	1–10	τρίγωνον ὀρθογώνιον right-angled triangle	right-angled triangle, linear dimension, feet → feet	inverse diagonal rule (P1i)
2	11–25	τρίγωνον ὀρθογώνιον right-angled triangle	right-angled triangle, linear dimension, feet → feet	inverse diagonal rule (P1i, implicitly)
3	26–40	τρίγωνον ὀρθογώνιον right-angled triangle	right-angled triangle, linear dimension, feet → feet	diagonal rule (P1, implicitly)

P.Lond. 5.1718

1	4.71	θησαυρός granary	parallelepiped, volume, units unclear	parallelepiped volume algorithm (S1)
2	4.72–73	πλοῖον ship	unclear	unclear
3	4.74	διῶρυξ trench	parallelepiped, volume, units unclear	presumably parallelepiped volume algorithm (S1)
4	4.75–76	λάκκος cistern	probably frustum, volume, units unclear	possibly conical frustum volume algorithm S3C
5	4.77–78	τοῖχος wall	parallelepiped?, volume, units unclear → bricks	

P.Mich. 3.145

1	3.2.1–9		numerical problem, artabas of wheat	
2	3.3.1–4		numerical problem, talents, drachmas of copper → staters, drachmas and obols of silver	

3	3.3.5–7		numerical problem, drachmas of copper → staters, drachmas and obols of silver	
4	3,3,8–11		numerical problem, talents, drachmas of silver → staters, drachmas and obols of copper	
5	3.4.1–18		numerical problem, apportionment of loads at various prices, drachmas	
6	3.5.1–4		numerical problem, freight charge on transport, stades, schoinoi, artabas → artabas	
7	3.5.5–8		numerical problem, freight charge on transport, artabas → artabas	
8	3.6.1–2		numerical problem, freight charge, artabas → artabas	
9	3.6.3–4		numerical problem, freight charge, artabas → artabas	
10	3.6.5–8	χωρίον property	rectangle, area, schoinia and arouras → schoinia	inverse rectangle area algorithm (P2i)
11	3.7.1–4		numerical problem, interest, drachmas and obols → drachmas	
12	3.7.5–7		numerical problem, interest, drachmas and obols → drachmas	
13	3.7.8–10		conversion of volume units, 36-choinix artabas → 40-choinix artabas	

P.Oxy. 3.470

1	31–46	ὡρολόγιον, ὀλμίσκος water-clock, bowl	conical surface, area, fingers → surface fingers	conic frustum surface algorithm (S4)

PSI 3.186

1	r.1–10	δέαδρον theater	conical surface?, area, number of seats → number of seats	
2	v.1–8	[ ] στρογγυλοῦν round [ ]	cylinder, volume, cubits → artabas	circle area algorithm A (P10A), prism volume algorithm S2

SB 16.12680 **verso**

1	v1–6	θησαυρός granary	unclear shape, volume, ? → artabas	
2	v7–10	[χωρίον] [property]	rectangle, linear dimension, schoinia and arouras → schoinia	inverse rectangle area algorithm (P2i)
3	v11–16	[χωρίον] [property]	rectangle, linear dimension, schoinia, arouras, and hammata → schoinia	inverse rectangle area algorithm (P2i)
4	v17–18		too fragmentary to identify	

T.Varie 20

1	1–2	συναβατερ *synabater* (?, λάκκος, cistern also mentioned, cf. Morgan Library tablet 1v1–6)	unclear shape, linear dimension?, cubits → some kind of feet (ποδ νομει)	
2	3–4	θησαυρός granary	parallelepiped, volume, cubits → solid feet and artabas	parallelepiped volume algorithm (S2)

3	5–6	θησαυρός granary	parallelepiped, volume, cubits → solid feet and artabas	parallelepiped volume algorithm (S2)
4	7–8bis	θησαυρός granary	parallelepiped?, linear dimension?, xyla and palms → some kind of feet (ποδ νομει)	

T. Varie 71–78

1	1v1–6	εἰδέαν βατέρου τοῦ λάκκου, form of a *bateros* (?) of the cistern (cf. T.Vat.gr. 62B, 1–2)	parallelepiped?, volume, cubits → naubia	
2	1v7–11	λάκκος στρογγύλον round cistern	frustum (actually cylinder), volume, cubits → naubia	
3	1v12–14	ἅμματα σκαφθέντα excavated hammata	prism of indeterminate cross-section, volume, hammata and fingers → naubia	
4	1v15–20	ναύβ(ιον) τετράκ(ωνον) rectangular trench	parallelepiped, volume, cubits → naubia	
5	2r1–5		prorating of solidi	
6	2r6–10	ναύβ(ιον) τετράκ(ωνον) rectangular trench	parallelepiped, volume, cubits → naubia	
7	2v1–5	ἅμματα σκαφθέντα excavated hammata	prism of indeterminate cross-section, volume, hammata and fingers → naubia	
8	2v6–7		damaged, involves cubits	
9	2v8–9		square, side in reeds	
10	3r1–5	ἅμματα σκαφθέντα excavated hammata	prism of indeterminate cross-section, volume, hammata and fingers → naubia	
11	3r6–7		conversion of cubits to fingers	
12	3r8–12	ναύβ(ιον) τετράκ(ωνον) rectangular trench	parallelepiped, volume, cubits → naubia	

13	3v1–7	διῶρυξ ποταμόν trench of a canal	parallelepiped?, volume, cubits → naubia	
14	3v8–12		prorating price of artabas in keratia	
15	4r1–4		too damaged to interpret	
16	4v1–4		rectangle, area, cubits and reeds → hammata	
17	4v5–9		quadrilateral, area, reeds → bikoi	
18	4v10		metrological statement?	

APPENDIX B. BIBLIOGRAPHIC CONCORDANCE OF MATHEMATICAL AND METROLOGICAL PAPYRI.

Checklist	Inventory and Other Designations	TM	Editions, Revisions, and Discussions
	BM Add. MS 41203A	65052	Skeat 1936 (edition)
	Chester Beatty Codex AC. 1390	61614	Brashear, Funk, Robinson, and Smith 1990 (edition)
	P.Berl. 11529v	63349	Schubart 1916 (edition), Boyaval 1978 (discussion)
	P.Col. inv. 157a	113807	Bakker 2007 (edition)
MPER NS 1.1	P.Vind. G 19996	63192	MPER NS 15.151 (edition of a table omitted in the original ed.)
MPER NS 15.172–174	P.Vind. G. 353a + 353 b+c + 256 + 59529 + 59530, SB 24.16273	63720	Bruins, Sijpesteijn, and Worp 1977 (prior ed.), Liesker and Sijpesteijn 1996 (revised text)
MPER NS 15.178	P.Vind. G 26740	60599	Bruins, Sijpesteijn, and Worp 1974.
P.Bagnall 35	P.Cornell inv. 69	65836	Bülow-Jacobsen and Taisbak 2003 (prior ed.)
P.Cair. cat. 10758	P.Akhmim, Akhmim Mathematical Papyrus	64999	Baillet 1892 (edition), Smyly 1920 (discussion)
P.Chic. 3	P.Ayer	63301	Goodspeed 1898 (prior ed.), Goodspeed 1903 (discussion), Boyaval 1978 (discussion)
P.Gen. 3.124	P.Gen. inv. 259	63474	Rudhardt 1978 (prior ed.), Sesiano 1999 (discussion)
P.Lond. 5.1718		65048	Skeat 1936 (revised text)
P.Mich. 3.145		63556	
P.Oxy. 3.470		63090	
P.Oxy. 4.669 verso		63999	
P.Oxy. 1.9+49.3456		64289	
P.Oxy. 49.3457		63295	
P.Oxy. 49.3458		64171	
P.Oxy. 49.3459		64172	
P.Oxy. 49.3460		63887	

P.Ryl. 2.64		64584	
PSI 3.186		64414	Smyly 1920 (revised text), Boyaval 1978 (revised text), Shelton 1981b (revised text)
SB 16.12680 verso	P.Ghent 1	63476	SB 3.6951 verso (prior ed.), Shelton 1981a (revised text)
T.Varie 20	P.Vat. gr. 62B	65105	
T.Varie 71–78	Morgan Library inv. 1032	65067	

References

Standard editions of papyri are cited according to the *Checklist of Editions of Greek, Latin, Demotic, and Coptic Papyri, Ostraca, and Tablets* at http://www.papyri.info/docs/checklist. See also the Bibliographical Concordance of Mathematical and Metrological Papyri.

Bagnall, Roger S. (1985) *Currency and Inflation in Fourth Century Egypt*. BASP Suppl. 5. Atlanta: Scholars Press.

Bagnall, Roger S. (1999) "The Date of P.Kell. I G. 62 and the Meaning of χωρίον," *Chronique d'Égypte* 74: 329–333.

Bagnall, Roger S., ed. (2009a) *The Oxford Handbook of Papyrology*. Oxford: Oxford University Press.

Bagnall, Roger S., ed. (2009b) "Practical Help: Chronology, Geography, Measures, Currency, Names, Prosopography, and Technical Vocabulary," in Bagnall 2009a, 179–196.

Bagnall, Roger S. and Bransbourg, Gilles (2019) "The Constantian Monetary Revolution," *ISAW Papers* 14; also to appear in T. Faucher, ed., *Money Rules*, Cairo: IFAO, forthcoming.

Bagnall, Roger S. and Worp, Klaas A. (2004) *Chronological Systems of Byzantine Egypt*, 2nd ed. Leiden: Brill.

Baillet, J. (1892) *Le papyrus mathématique d'Akhmîm*, Mémoires publiés par les membres de la Mission archéologique française au Caire 9.1. Paris: Leroux.

Bakker, Marja (2007) "A Papyrus with Mathematical Problems," *BASP* 44: 7–21.

Benaissa, Amin (2012) *Rural Settlements of the Oxyrhynchite Nome* version 2.0. Leuven: Trismegistos Online Publications. https://www.trismegistos.org/top.php.

Blume, F., K. Lachmann, and A. Rudorff (1848–1852) *Die Schriften der römischen Feldmesser*, 2 vols., Berlin: Georg Reimer.

Boozer, Anna (2015) "Building Domestic Space: The Construction Techniques for House B2," in Anna Boozer, ed., *Amheida II: A Late Romano-Egyptian House in the Dakhla Oasis, Amheida House B2*. New York: New York University Press and Institute for the Study of the Ancient World, 141–155.

Boyaval, Bernard (1978) "Notes métrologiques," *ZPE* 28: 203–215.

Brashear, W. M., Funk, W.-P., Robinson, J. M., and Smith, R. (1990) *The Chester Beatty Codex AC. 1390. Mathematical school exercises in Greek and John 10:7–13:38 in Subachmimic*, Chester Beatty Monographs 13. Leuven: Peeters.

Britton, J. P., Proust, C., and Shnider, S. (2011) "Plimpton 322: A Review and a Different Perspective," *Archive for History of Exact Sciences* 65: 519–566.

Bruins, Evert M., P. J. Sijpesteijn, and K. Worp (1974) "A Greek Mathematical Papyrus," *Janus* 61: 297–312.

Bruins, Evert M., P. J. Sijpesteijn, and K. Worp (1977) "Fragments of Mathematics on Papyrus," *Chronique d'Égypte* 52: 105–111.

Bülow-Jacobsen, Adam, and C. M. Taisbak (2003) "P.Cornell inv. 69: Fragment of a Handbook in Geometry," in Anders Piltz *et al.*, eds., *For Particular Reasons: Studies in Honour of Jerker Blomqvist*, Lund: Nordic Academic Press, 54–70.

Cribiore, Raffaella (1996) *Writing, Teachers, and Students in Graeco-Roman Egypt*. Atlanta: Scholars Press.

Cribiore, Raffaella (2001) *Gymnastics of the Mind. Greek Education in Hellenistic and Roman Egypt*. Princeton: Princeton University Press.

Emmel, Stephen (2015) "On the Discovery of the Gospel of Judas Codex ('Codex Tchacos') 1982–1983." Internet publication dated 2015-07-27. https://www.academia.edu/15353011/ On_the_Discovery_of_the_Gospel_of_Judas_Codex_Codex_Tchacos_1982_1983_2006_

Finckh, H. (1962) *Das Zinsrecht der gräko-ägyptischen Papyri*. Diss. Erlangen.

Friberg, Jöran (2007) *A Remarkable Collection of Babylonian Mathematical Texts*. New York: Springer.

Gignac, Francis T. (1976–1981) *A Grammar of the Greek Papyri of the Roman and Byzantine Periods*, 2 vols., Milan: Istituto editoriale Cisalpino-La Goliardica.

Goodspeed, Edgar J. (1898) "The Ayer Papyrus: A Mathematical Fragment," *American Journal of Philology* 19.1: 25–39.

Goodspeed, Edgar J. (1903) "The Ayer Papyrus," *American Mathematical Monthly* 10.5: 133–135.

Herrmann, J. (1958) *Studien zur Bodenpacht im Recht der graeco-aegyptischen Papyri*. Munich: Beck.

Husselman, Elinor M. (1952) "The Granaries of Karanis," *Transactions and Proceedings of the American Philological Association* 83: 56–73.

Johnson, Allan Chester (1936) *Roman Egypt to the Reign of Diocletian*. An Economic Survey of Ancient Rome, ed. Tenney Frank, vol. 2. Baltimore: The Johns Hopkins Press.

Kasser, R., Meyer, M., Wurst, G., and Gaudard, F. (2007) *The Lost Gospel of Judas together with the Letter of Peter to Philip, James, and a Book of Allogenes from the Codex Tchacos: Critical Edition*. Washington, DC: National Geographic.

Keenan, James G. (1973) (1974) "The Names Flavius and Aurelius as Status Designations in Later Roman Egypt," *ZPE* 11: 33–63, 13: 283–304.

Keenan, J. G. (1983) "An Afterthought on the Names Flavius and Aurelius," *ZPE* 53: 245–250.

Krosney, Herbert (2006) *The Lost Gospel: The Quest for the Gospel of Judas Iscariot*. Washington, DC: National Geographic.

Liesker, W. H. M. and Sijpesteijn, P. J. (1996) "Bruchstücke antiker Geometrie," *ZPE* 113: 183–186.

Mandilaras, Basil G. (1973) *The Verb in the Greek Non-Literary Papyri*, Athens: Ministry of Culture and Sciences.

Minutoli, Diletta and Pintaudi, Rosario (2011) "Un codice biblico su papiro della collezione Schøyen MS 187 (*Esodo* IV 16 – VII 21)," in G. Bastianini and A. Casanova, eds., *I papiri letterari cristiani: Atti del convegno internazionale di studi in memoria di Mario Naldini Firenze, 10–11 giugno 2010*. Florence: Istituto Papirologico «G. Vitelli», 193–205.

Morgan, D. P. and Chemla, K. (2018) "Writing in Turns: An Analysis of Scribal Hands in the Bamboo Manuscript *Suan shu shu* 筭數書 (*Writings on Mathematical Procedures*) from Zhangjiashan Tomb No. 247," *Bamboo and Silk* 1: 152–190.

Nongbri, Brent (2018) *God's Library: The Archaeology of the Earliest Christian Manuscripts*. New Haven: Yale University Press.

Parker, Richard A. (1972) *Demotic Mathematical Papyri*, Providence: Brown University Press.

Parker, Richard A. (1975) "A Mathematical Exercise: P. Dem. Heidelberg 663," *Journal of Egyptian Archaeology* 61: 189–196.

Pottage, John (1974) "The Mensuration of Quadrilaterals and the Generation of Pythagorean Triads: A Mathematical, Heuristic and Historical Study with Special Reference to Brahmagupta's Rules," *Archive for History of Exact Sciences* 14.4: 299–354.

Pruneti, Paola (1981) *I centri abitati dell'Ossirinchite*, Pap.Flor. 9. Florence: Gonnelli.

Rathbone, Dominic W. (1991) *Economic rationalism and rural society in third-century A.D. Egypt*. Cambridge: Cambridge University Press.

Rossi, Corinna and Fausta Fiorillo (2018) "A Metrological Study of the Late Roman Fort of Umm al-Dabadib, Kharga Oasis (Egypt)," *Nexus Network Journal* 20: 373–391.

Rowlandson, Jane (1996) *Landowners and Tenants in Roman Egypt*. Oxford: Oxford University Press.

Rudhardt, Jean (1978) "Trois problèmes de géométrie, conservés par un papyrus genevois," *Museum Helveticum* 35: 233–240 and plate 7.

Schnebel, Michael (1925) *Die Landwirtschaft im hellenistischen Ägypten*, vol. 1: *Der Betrieb der Landwirtschaft*. Munich: C. H. Beck.

Schubart, Wilhem (1916) "Mathematische Aufgaben auf Papyrus," *Amtliche Berichte aus den Königlichen Kunstsammlungen* 37.8: 5–9.

Sesiano, Jacques (1999) "Sur le Papyrus graecus genevensis 259," *Museum Helveticum* 56: 26–32.

Shelton, John C. (1981a) "Mathematical Problems on a Papyrus from the Gent Collection (SB III 6951 verso)," *ZPE* 42: 91–94.

Shelton, John C. (1981b) "Two Notes on the Artab," *ZPE* 42: 99–106.

Skeat, T. C. (1936) "A Greek Mathematical Tablet," *Mizraim* 3: 18–25.

Smyly, J. Gilbart (1920) "Some Examples of Greek Arithmetic," *Hermathena* 19: 105–114.

Spencer, A. J. (1979) *Brick Architecture in Ancient Egypt*. Warminster: Aris & Phillips.

Struve, W. W. and Turajeff, B. A. (1930) *Mathematischer Papyrus des Staatlichen Museums der schönen Künste in Moskau*, Quellen und Studien zur Geschichte der Mathematik A 1. Berlin: Springer.

Tou, Erik (2014) "Measuring the Accuracy of an Ancient Area Formula," SIAS Faculty Publications. 848. https://digitalcommons.tacoma.uw.edu/ias_pub/848 .

Turner, E. G. (1977) *The Typology of the Early Codex*. Philadelphia: University of Pennsylvania Press.

Youtie, Herbert C. (1976) "P. Mich. Inv. 406: Loan of Money with Interest in Kind," *ZPE* 23: 139–142.

Index

1. Index to the mathematical problems and metrological texts

The index omits numerals, the definite article, letters not identifiable as part of a word, γίνομαι, εἰμί, and καί.

ἄγρος Ev21–22, Ev23, Mr13
Αἰγύπτιος Gr12
ἄκαινα Gr3
ἀλλήλων Ov7
ἄλλος Cv2, Dr3, Dv2, Fv11, Gv2, Mv3, Mv6, Mv7
ἄλλως Gr9
ἄμμα Gr3, [H →0], H →2, H →16
ἀμπέλιον Fv18
ἄμπελος Fv20, Fv22
ἄμφοδον Gr6, Mr6, Mr8
ἀνά Av11, Cv1, Dr2, Gv2, Ov17
ἀναβάλλω Cr14, Er13, Nv2, Nv8
ἀναλύω Dv21, Er14
ἀντιλαμβάνω: ἀντιλήψεται Dr15, Dr19
ἄνω Dv19, Er2, Er18,
ἀπηλιώτης Br6, Fv1, Fv6, Fv19, Or9, Ov2, Ov6
ἀπό Av13, Av14, [Br5], [Br6], [Bv2], Cr14, Dr6, Dr17, [Gv5], Er6, Fv14, Fv19, Mv4, Or11
ἀπόδειξις Dv5, Fr2, Ov7
ἄρα Br21, Cr20, Cv5, Dr7, Dr9, Dr12, Dr18, Dv8, Dv14, [Er1], [Gv6], [Gv8], [Gv9], Nr18, Er17, Fv15, Fv21, I →22, Mv6, Mv8, Mv9, Mv10, Mv12, Nr5, Nr8, Nr18, {Nr18}, Or6, Or19, [Ov19]
ἀργύριον Ov13

ἄρουρα Dr10, Dv11, Ev18, Ev21, Ev22, Ev23, Fv2, Fv9, Gv8, Mr13, Or10, Ov3, Or9, I ↓18?

ἀρτάβη Cr17, Cr20, Ev17, Fr1, Fr4, Fr4, Fr5, Fr7, Fr8, Fr9, Fr10, Mr12, Mv7, Mv8, Mv10, Mv11, Mv13, Mv13, Nr1, Nr6, Nr12, Nr19, Nv13, Or3, Or6

αὐτός Br8, Dr12, Gr14, Gr17

αὐτοῦ Cv4

βάθος Ar16, Ar18, Cv14, Cv17–18, Dv20, Dv23, Er3, Er7, Er12, Er16, Ev6, Ev14, Gr23, Gr26, [I →16], [I →18], Mr10, Nr11, Nr16, [Nv1], Nv4, Nv6, Nv13, Nv15, Or2, Or4

βαρβαρικός Mr3

βασιλικός Ev8–9, Ev12

βάσις Av12, Av13, Av15, Av18, Bv3, [Cv1–2], Cv4, Cv7, Dr3, Dr4–5, Dr8, Dr12, [Dv3–4], Dv9, Dv10, Dv13, Dv15, Fv15, Fv17, Gv2, Gv4, Gv6, Gv9

βῆμα Fv1, Fv1, Fv5, [Gr2], [H →0], H →3, H →5, H →7, H →9, H →17

βῖκος Ev19, Ev19, Ev22, Ev22, Mr14, Or15, Or18, Or20

βορέας Ov4; βορρᾶς Br5, Bv3, Fv1, Fv3, Fv18, Or8, Ov1

γάρ Gr12

γεωμετρικός Ev1, Ev3, Ev12, [H →-1], Mr16, Mr17, Mr18–19

γράμμα [I →8], I →9, [I →9]

γύης Mr15, Mr15, Mr18

δάκτυλος Av1, Av2, Av4, [Av5], Av6, Av7 (2x), Bv9, Bv10, Bv14, Bv15, Bv16, Cr2, Cr3, Cr3, Cr4, Cr10, Cr10, Ev11, Ev9, Gr1, Gr7, Gr13, Gr13–Gr14, Gv11, Gv11, Gv15, Gv15–16, Gv17, H →2, H →4, H →6, H →8, H →10, H →12, H →13, H →14, H →15, H →16, H →19, [I →1], I →2

δέ Ar15, Br11, Br12, Bv9, Cr2, Cr8, [Ev1], Ev5, Ev9, Ev10, Ev11, Ev11, Ev13, Ev16, Ev17, Ev18, Ev21, Fv19, Gr13, Gr14, Gr16, Gr17, H →2, H →5, [H →7], [H →9], H →10, H →12, H →13, H →14, H →15, [I →1], I →3, [I →5], [I →6], I →7, I →8, [I →8], I →9, I →10, I →11, I →11, I →12, I →13, I →14, I →20, Mr7, Mr9, Mr10, Mr13, Mr14, Mr15, Mr15, Mr16, [Mr16], Nr10, Nr11, [Ov15]

δέκα. Ar18, Or23

δέκατος Fr2, Fr3

δεύτερος Dr15, Dr18, Fr8, Fr8, Fr14, Mv8

δημόσιος Cr15, Ev15, Gr17, Mr11

διά Dv6, Fv4, [I →19], Or18

διαιρέω Gr14

διάκοπος Ar14

διάμετρος Dv19, Er1, Er2, Er4, Nv1, Nv3

διάστασις Gr20, Nr9

δίαυλον Gr4

διότι Nv7
διπλόω Or10, Or10
δισχίλιοι [I →14]
διῶρυξ Cv13
δόλιχος Gr4
δραχμή I →11, I →11, I →13
δύο Fv13, Mr8, Nv3
δωδεκατικός Gr5, Mr5, Mr8
δωδέκατος Cr16, Cr18

ἑαυτοῦ: ἑαυτά Av16, Av17, Br7, Bv5, Cv3, Dr5, [Dr5], Er4–5, Fv13, Fv13, Gr25, Gv4, Gv4, Mv3, Nv5
εἰ [Ov15]
εἶδος Gr1
εἰς Av5, Br4, Br5, Br6, Bv2, [Bv3], Bv14, Cr10, Cr14, Cr15, Cr16, Dr13, Dv21, Er15, Fv13, Fv19, Gr14, Gv15
εἷς, μία, ἕν Cr19, Ev13, Ev14, Mr10, Mr10, Or24, Or25
εἰσέρχομαι Fr9
ἐκ Ev13, Mr9
ἕκαστος Av16, Cv7, Dr7, Fr11, Gv6, Gv9
ἑκατοστή Fr11, Fr12
ἐλάχιστος Gr7–Gr8
ἐμβαδικός Ev5
ἐμβαδόν Av19, Av20, Br1, Br8, Bv4, Cv6, Dr8, Dr9–10, Dv9, Dv11, Gv6, Gv8
ἐμβαδός Ev7, Ev19–20, Ev21, Ev22–23, Ev24, Mr7, Or19
ἐν Av8, Cv11, Dr15, Dr19, Ev8, Ev18, Ev20, Fr19, Fr22, H →21, I ↓13
ἑξακισχίλιοι [I →14]
ἐπανέρχομαι Dr14
ἐπειδή Av12–13
ἐπί Ar14, Ar18, Ar19, Av3, [Av3], Av4, Av16, Av17, Av19, Av21, Br7, Br16, [Br16], Br18, Br18, [Br19], Br20, [Bv5], Bv12, Bv13, Bv5, Cr8, Cr8, Cr19, Cv3, Cv4, Cv7, [Cv7], Cv15, Cv16, Cv17, Cv17, Dr5, Dr5, Dr8, [Dr8], Dr16, Dv6, Dv6, Dv7, Dv8, Dv8, Dv10, Dv10, Dv20, Dv22, [Er4], Er7, Er15, Er16, Fr2, Fr13, Fr14, Fr15, Fv4, Fv4, Fv7, Fv7, Fv13, Fv13, Fv18, Fv20, Gr25, Gr26, Gv4, Gv4, [Gv7], Gv7, Gv12, Gv13, Gv13, [Gv14], I →18, I →18, [I →18], I →19, [I →19], I →20, I →21, [I →21], [Mv3], [Mv4], Nr3, Nr3, Nr4, Nr5, Nr6, Nr14, Nr14, Nr15, Nr16, Nr17, Nr17, Nv5, Nv6, Nv14, Nv14, Nv15, Nv16, Or4, Or4, Or4, Or5, Or5, Or9, Or16, Or16, [Or23], Ov7, Ov8, [Ov9], Ov9, Ov12, Ov14, Ov16, Ov16,
ἐργάζομαι Ov14
ἐργάτης Ov14

ἐσώτερος Br11

ἕτερος Dr14

εὐθυμετρικός Ev3, Ev4

εὑρίσκω: εὑρεῖν Av19, Av20, Br13, Bv4, Cr4, Cv2, Cv6, Dr3, Dr7, Dr15, Dv2, Dv9, Dv11, Er3, Er13, Fr10, Fv2, Fv11, Fv20, Gr23, Gv2, Gv6, Gv11, Nr1, Nr12, Nv2, Nv13, Or3, Or15, Ov3, I →17

ἔχω Ar20, Av6, Br2, [Br4], Br8, Br10, Br11, Bv2, Bv6, Bv15, Bv16, Cr11, Cr20, Cv8, Cv15, Cv16, Cv19, Dr10, Dv16, Dv22, Er9, Er18, [Ev1], Ev13, Ev14, Ev19, Ev2, Ev22, Ev24, Ev7, Fr4, Fr7, Fr8, Fr9, Fr11, Fr11, Fr12, Fr16, Fr18, Fv5, Fv9, Fv16, Fv22, Gr27, Gv8, Gv16, [H →0], H →3, H →5, H →7, H →9, [H →11], H →12, [H →14], [H →15], [H →16], H →16, [I →0], I →1, [I →2], I →3, [I →4], I →5, I →6, [I →7], I →8, I →9, I →10, [I →10], I →11, I →23, Mr6, Mr10, Mr14, Mr14, Mr15, Mr15, Mr16, <Mr17>, [Mr17], Mv14, [Nv7], Nv16, [Or6], Or12 , Or19 , Or20, Or26, Ov10, Ov19

ἤ Ev6

ἡμέρα Dr13, Dr13, Dr15, Dr19, Ov12, Ov14

ἡμιούγκιον I →8, I →9

ἥμισυ Ar18, Av3, Av15, Av20, Br15, Bv12, Cr6, Cr7, Cr15, Cr17–18, Cv3, Cv8, Dr4, Dr9, Dv10, Fv4, Fv7, Gr24, Gv3, Gv7, Mv3, [Nr3], Nv4, Ov5, Ov7, Ov7

ἥτις Gr8

ἤτοι Ev6

θερμός I →5, I →6, I →7

θησαυρός Cr14, Fr7, Fr7, Fr8, Fr9, Nr1–2, Nr5, Nr8, Nr12–13, Nr18, Nv12, Or1, Or6

ἴδιος Gr9

ἰδιωτικός Ev10, Ev12, Ev16

ἴσος Gv10

ἰσοσκελής Av11, Cv1, Dr2, Gv1, [Gv2]

*ἱστονικός Gr19

Ἰταλικός Gr12

καθάπερ Gr8

κάλαμος Gr3, [H →0], H →3, H →5, H →17

καλέω Gr6, Gr16, Gr17, Mr4, Mr5, Mr8

καμάρα Nr11, Nr15, Or2–3, Or5

καμαρωτός Nr8, Or1

κατά Ev18, Ev21, Ev23, Ev5, Ev6, Mr13, Mr14

καταλείπω Cr16

κάτω Ar15, Nv1

μύριος Ev21

ναύβιον Ar20, Cv18, Dv20, Dv21, Dv22, Er3–4, Er7, Er8, Er13, Er14, Ev8, Ev13, Ev15, Gr22, H →16, I ↓16, Mr9, Mr11, Nv2, Nv7–8, Nv9
νέος Cr1
νιλομετρικός Gr18
νότος Br5, [Bv2], [Fv1], Fv3, Fv18, Or8, Or12 , [Ov1], Ov4

ξέστης I →17, [I →23]
ξηρός Ev17, Mr12
ξύλον Cr1, Cr5, Cv14, Cv16, Cv20, Ev8, Ev12, Ev13, Ev14, Ev14, Gr3, [H →0], H →3, H →5, H →7, H →17, Mr10, Mr10, Mr10
ξυλοτομέω: ξυλοτομοῦντος Ov12

ὀβολός I →14
ὄγδοος Ev1, [Ev1–2]
ὅδε Gr1, Gr6–7
οἰκοπεδικός Ev7
οἰκόπεδον Ev18, Ev20
ὀκταπλόω Or22–23
ὀκτωκαιδέκατος Dr18
ὅμοιος Nr6
ὁμοίως Av6, Br2, Br17, Bv6, Bv15, Cr11, Cr21, Cv9, Cv19, Dr10–11, Dv16, Er9, Er19, Fr14, Fr15, Fr16, Fv9, Fv16, Fv22, Gr27, Gv8, Gv16, I →23, Mv14, Nr15, [Or6], Or26, Ov19
ὀνηλάτης Cr16
ὄργυια Gr3, Gr6
ὀρθός Cv7, Dr7, Dr8, Dv3, Dv7, Dv7, Dv10, Dv13, Dv14, Fv15, Fv17, Gv6, Gv7, Gv9
ὀρθογώνιος Av15, Av16, Dv3, Fv11
ὄρυγμα Dv19, Er2, Er11
ὅς, ἥ, ὅ Ar14, Ar18, Av3, Av11, Av15, Av18, Av20, Br4, Br15, Bv2, Bv12, Cr6, Cr7, Cv5, Cv8, [Dr7], Dr9, Dv10, Dv25, Er11, Er17, Ev8, Fv4, Fv6, Fv14, [Gr5], Gv5, Gv7, [Mr4], Mv3, Nr3, Nv4, Or24, Ov5, [Ov7]
ὁσοσδήποτε Br5, Dv20, Er12, Or8
ὅτι [Dv6], [Fv4], I →20, Or18
οὐ, οὐκ Gr8
οὐγκία [I →7], I →8
οὗτος Cr9, Dv24, Er5, Fv21, Gr12, Gr25, Gv13, Gv14, Nr4, Nr16, Nv5, Nv7, Nv14, Or17, Or25
οὕτω, οὕτως Ar17, Ar20, Av2, Av6, [Br1], [Br7], Br8, Br14, Bv4, Bv6, Bv10, Bv15, Bv16, Cr5, Cr11, Cr17, Cr20, Cv6, Cv8, Cv14, Cv19, [Dr3–4], Dr10, Dr16, Dv16, Dv2, Dv21,

Er14, Er18, Er4, Er8, Ev24, Fr4, Fr10, Fr16, Fr18, Fv3, Fv9, Fv12, Fv16, Fv22, Gr24, Gr26–27, Gv3, Gv8, Gv12, Gv16, H →19, I →23, Mv2, Mv14, Nr2, Nr13, Nv3, Nv13, Nv16, Or3, Or6, Or9, Or12 , Or16, Or20, Or22, [Or25], Ov3, Ov10, Ov19

παλαιστής Ev9, Ev10–11, [Gr1], Gr10, Gr10, Gr10–11, Gr11, Gr15, Gr16, Gr18, Gr18–Gr19, Gr19, H →10, H →11, H →2, H →4, H →6, H →8, H →13, H →14, H →15, H →15, H →18, I →1, [I →1]

πάλιν Bv12, Ov5, Ov8

πανταχόθεν Mr9

παρά Ar20, [Av5], Bv5, Bv14, Cr9, Cv18, Dv6, Dv23, Er8, Er15, Fr13, <Fv8>, Gv14, Mv5, Nv7, Or11, Or18, Or18, [Ov10], Ov18

παρασάγγης Gr5

πάχος [Av2], Av4, Av7, [Br11], Br16, Bv9, Bv13, Cr3, Cr7, Ev6, Gv11, Gv13

πέμπτος Mv12

πέντε Fr22, Gr15, Mv3

περί Br10

περιοχή Dv4, Dv12, Dv13, Dv15

περιφέρεια Av7, Er11, Er13–14, Er17, Er18, [Gr22], Gr24–25

πῆχυς Ar15, Ar16, Ar16, Ar16, Ar18–19, Ar19, Av1, Av7, Br11, Br11, Br12, Br12, Br13, Bv13, Bv16, Bv8, Bv9, Cr1, Cr8, Cv14, Cv15, Cv16, Cv18, Dv19, Dv20, Dv21, Dv22, Dv23, Er1, Er3, Er3, Er7, Er12, Er12, Er15, Er16, Er18, Ev2, Ev3, Ev4, Ev4, Ev7, Ev7, Ev9, Ev10, Ev14, Ev15, Ev16, Ev16, Ev20, Ev21, Ev23, Ev24, Fv2, Fv5, Fv5, Fv7, Fv18– 19, Fv19, Gr2, Gr15, Gr16, Gr18, Gr19, Gr23, Gr23, Gr26, Gv10, Gv17, H →1, H →3 , H →5, H →7, H →9, H →10, H →17, [I →0], I →2, I →16, [I →16], I →17, I →19, Mr6, Mr6, Mr7, Mr7, Mr10, Mr12, Mr12, Mr17, Mr19, Nr4, Nr10, Nr10, Nr11, Nr11–12, Nv1, Nv2, Nv6, Nv8, Nv12, Nv12, Nv13, Nv15, Or1, Or2, Or2, Or3, Or14, Or14, Or15, Or17, Or19

πιπράσκω: πραθήσονται Fr3

πλάτος Ar16, Ar17, Av1, Av3, Av7, Br13, Bv9, Bv11, Cr6, Cv13, Ev6, [Gv11], Gv12, I →16, I →18, Mr7, Nr3, Nr10, Nr14, Nv12, Nv14, [Or1–2], Or4, [Or4], Or14

πλέθρον [Gr3]

πλευρά Av18, Cv2, Cv5, [Cv5], Dr3, Dr7, Dv2, Dv25, Er17, Fv12, Fv13, Fv15, Gv3, Gv5, Or22, Or24

πλευρόν Br4, Bv2

πλίνθος Br21, Br14

πλοῖον Cr14, Cr16–17, Cr20

ποιέω: ποιοῦμαι Ar17, Br14–15, Br7, Cr5, Dr4, Dv21, I →17

– ποιοῦμεν Av2, Bv10, Cr17, Cv6, Cv14, Dr16, Dv2, Er4, Er14, Fr10, Fv3, Fv12, Gr24, Gv3, Gv12, Mv2, Nr2, Nv13, [Or3], Or10, Or16, Or22, Ov3

– ποιῶ Nr4, Nr13, Nv3

πόλις Ev18, Mr14

πολυπλασιάζω Bv12, Nr2, Nr13, Nr15, Nr16–17,
πόσος Br14, Dr15, Fr2, Fr10, Fr11, Fr12, Fv20, [Gr23], Nr1, Nr12, Nv2, Or15, Ov15
ποταμίτης Ov15
ποταμός Cv13
πούς Gr2, Gr11, Gr12, Gr13, H →1, H →3, [H →5], H →7, H →9, H →11, H →12, [H →18]
προάγω Gr7
προβαίνω Av8, Cv11, Fr20, Fr23, H →21–H →22
προστίθημι: προσθές Or23
– προστίθομεν Dv24, Fr3, Mv6, Mv7, Mv9, Mv10, Mv11–12
πρῶτον Dr16, Dr19
πρῶτος Fr7, Mv6
πτολεμαικός Gr11
πυγών Gr2, Gr16
πυθαγορικός Dv3
πύργος Br10, Br21
πυρός Cr17, Cr20, Fr1, Mv7, Mv8, Mv10, Mv11, Mv12
πωλέω Fr1, Ov17

σῖτος Nr6
σκέλος Av11, Gr20
σπιθαμή see ψιθαμή
στάδιον Dr13, Dr14, Gr4, [Mr15], Mr16, Mr18
στατήρ [I →10], I →11, [I →12]
στερεός Ev15, Ev17, Gv12, Mr11, Mr12
στρογγύλος Dv19, Er11, Er2, Gr22, Gv10
σύγκειμαι Gr9
σύν Dv4, Dv15, Dv15, Dv15
συντίθημι: συντίθω Ar17, [Av2], Br15, Br18, Br19, Bv10–11, Cr5, Cr17, Dv4, Dv12, Dv13,
    Dv16, Dv24, Fr12, Fr15, Fv3, Fv5, Mv13, Nv3
– συντίθομεν Cr6–7, Ov3–4, Ov5–6
σφραγίς Or8, I ↓18
σχῆμα Nr8
σχοινίον Av11–12, Br5, Br6, Bv3, [Bv3–4], Cv1, Cv3, Cv13, Cv15, Cv20, Dr2, Ev1, Ev2,
    Ev11, Gr4, Gv2, [H →-1], Mr2, Mr4, Mr4, Mr16, Mr16–17, Mr18, Or8, Ov1, Ov1, Ov2,
    Ov2
σχοῖνος Gr5, Mr3, Mr14, Mr18

τάλαντον Fr1, I →10, I →12, Ov13, Ov17, Ov19
τεκτονικός Gr13, Gr16–17
τέσσάρες I ↓13, Nr9

τέταρτος [I →4–5], I →5, Er5, Fv21, Mv11, Nv15
τετράγωνος Av18, Br1, Br13, Br17, Br4, Bv2, Bv8, Ev13
τετρακισμύριοι I →14
τετραορθογώνιος Av20–21
τετρόβολος I →12
τίς: τί Cr18, Dv6, Fv4, [I →19], Or18
τις Ar14, Cr14, Dr13, Fr9, [Fr10]
τουτέστιν Nr14, Nv3–4
τραπέζιον Av11, Bv8
τρεῖς Fr7
τρέχω Dr13, Dr14
τριακόσιοι Fr11
τρίγωνος [Cv1], Nv12, Or22
τρίτος Cr15, Cr18, Dv24, Fr9, Fr15, Mv9, Or25

ὑγρός Ev17, I →20, Mr13
ὑπέρ Cr15, Ov13, Ov15
ὑπερέχω Mv1
ὑπό Nr9
ὑποτείνω Dv4, Dv9, Dv14, Dv15, Fv11, Fv12, Fv15, Fv17
ὑφαιρέω: ὑφέλομαι Dr17
– ὑφέλομεν Av14, Dr6, Er5, Er6, Fv14, Gv5, Mv4, Or11, Or24
– ὑφέλω Av13, Nv5
ὕψος Br12, Br16, Ev6

φύλλον Cr4

χοῦς I →3, I →3
χῶμα Ar14, Ar15
χωρέω Br21, Cr20, Ev17, Fv20, Gr24, I →22, Mr12, Nr1, Nr12, Nr18, {Nr18}, Nr5, Or6
χωρίον Br4, Fv18, Fv21

ψιθαμή (σπιθαμή) Gr11, Gr2, H →1, H →4, H →6, H →8, H →9, H →11, H →12, H →13,
    H →18
ψιλός Or14, Or20

ὥστε Ev2, Ev11, Ev14, Ev20, Ev23, [I →1], [I →3], [I →6], [I →9], I →12, Mr11, Mr17

2. Index to the Model Documents

A. Chronology

ἔτος Ar6
   [ἔτος μα// καὶ] ι// H ↓4–5
   ἔτους .[(μ[---]?) I ↓8
Θώθ Ar6
Μεχείρ I ↓8

B. Personal Names

Ἀτρῆς I ↓3
Διογένης H ↓1
Φλάουιος H ↓1

C. Geographical Names

ἐποίκιον Ἱερέων H ↓3
Ἱερέων, ἐποίκιον H ↓3
κώμη I ↓3
νομός H ↓3; I ↓4
Ὀξυρυγχιτῶν πόλις H ↓2
πόλις H ↓2
πάγος H ↓3
χωρίον H ↓6
ϛ πάγος H ↓3

D. Money

ἀργύριον Ar7, I ↓6
δηνάριον I ↓6
δίζῳδος Ar4
εὐχάρακτος Ar3–4
νομισμάτιον Ar3
τάλαντον Ar7
χρυσός Ar3

E. Measures

ἄρουρα H ↓7, [7–8], 8, 9
ἀρτάβη H ↓11
δεσμίδιον H ↓10
μνᾶ I ↓5
πενταμναῖος H ↓10

F. General Index

αἱρέω Ar8
ἀκίνδυνος H ↓13
ἀκοιλάντως Ar7
ἀνά H ↓11
ἄνευ [Ar9]; [I ↓9]
ἀντί Ar4
ἀνυπερθέτως H ↓15.
ἁπλοῦς Ar11; I ↓11
ἀπό [H ↓1]; H ↓3; H ↓4; H ↓5; I ↓3; I ↓5
ἀποδίδωμι Ar8; H ↓14; I ↓7
ἀπότακτος Ar5
αὐτός H ↓3
βορρᾶς H ↓6
γίνομαι Ar9; I ↓9
γραμμάτιον Ar11
γράφω Ar11; I ↓11
δύο H ↓9,
δώδεκα H ↓10
ἐγώ Ar4; Ar10; Ar11; I ↓10 (2x)
ἔδαφος Ar6
εἰς Ar5
εἷς Ar4; H ↓8
ἐκ, ἐξ Ar3; Ar10; H ↓6; I ↓10 (2x)
ἑκουσίως H ↓3–4
ἐμαυτοῦ I ↓5
ἐν [Ar3]; H ↓7, 8, [8], 11, 12; I ↓2
ἐνεστῶς Ar6; I ↓8
ἐπάναγκες H ↓14; I ↓7

Plates

Plate I

A (recto)

Plate II

A (verso)

Plate III

B (recto)

Plate IV

B (verso)

Plate V

C (recto)

Plate VI

C (verso)

Plate VII

D (recto)

Plate VIII

D (verso)

Plate IX

E (recto)

Plate X

E (verso)

Plate XI

F (recto)

Plate XII

F (verso)

Plate XIII

G (recto)

Plate XIV

G (verso)

Plate XV

H (recto)

Plate XVI

H (verso)

Plate XVII

I (recto)

Plate XVIII

I (verso)

Plate XIX

M (recto)

Plate XX

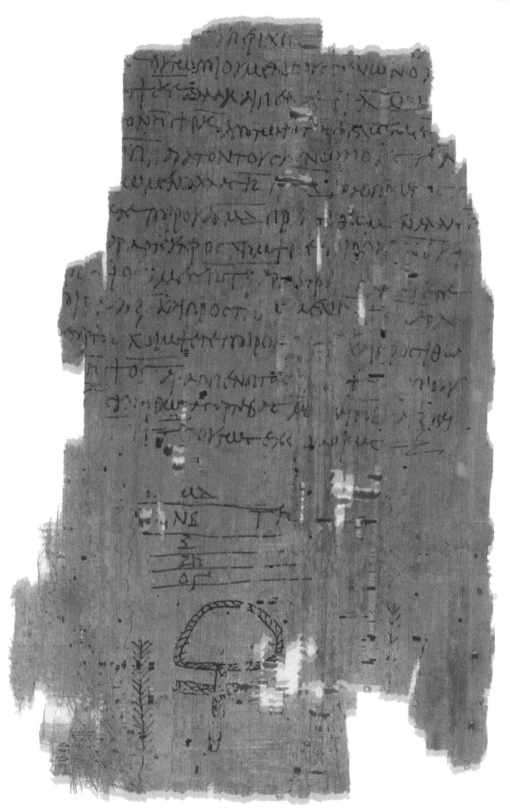

M (verso)

Plate XXI

N (recto)

Plate XXII

N (verso)

Plate XXIII

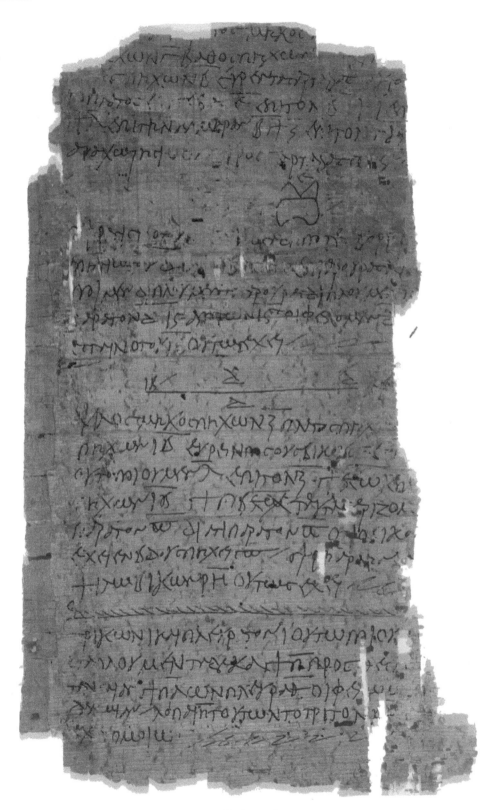

O (recto)

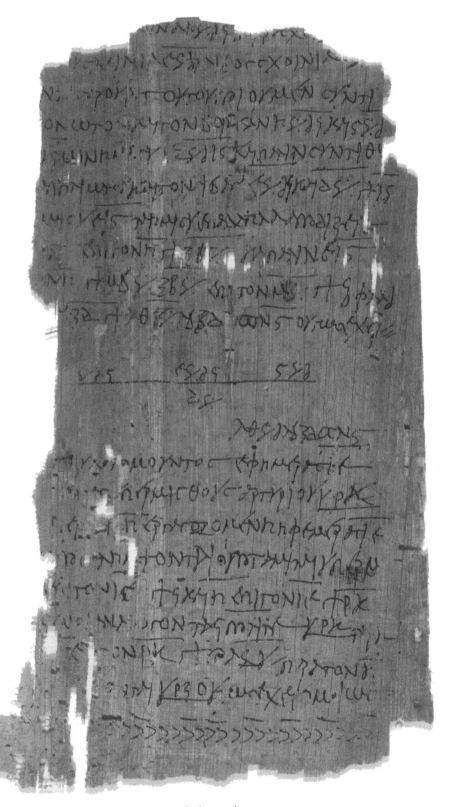

O (verso)

Plate XXIV

9 781479 801763